ENVIRONMENTAL ETHICS

About the Editor

Prof. R P Misra is a well-known scholar of development studies, urban and regional planning, and Gandhian Studies. Among the positions he has held are Professor of Geography and Founder Director, Institute of Development Studies, University of Mysore, Vice-Director, United Nations Centre for Regional Development, Nagoya, Japan, and Vice-Chancellor, Allahabad University, Allahabad. He has been Visiting Professor at several universities in India and abroad.

He is the founder-president of two non-governmental non-profit organizations: Delhi Society of Non-violence and Sustainable Development Foundation. He has worked in a number of countries of Asia, Africa and Latin America as United Nations Expert.

Prof. Misra has written and edited over forty books on Planning, Development, Ecology, Peace, Conflict Resolution, Education, Urbanization and Rural Development and has been a frequent contributor to professional journals. Currently he is engaged in editing a ten-volumes series on *Rediscovering Gandhi* and finalizing volumes on *Natural Principles of Regional Planning*, and *Million Cities of India*. He is actively involved in peace research and environmental movement in the country.

Prof. and Chairman, Centre for Gandhian Studies and Peace Research, Misra is currently professor of Geography, Delhi School of Economics, Delhi University, Delhi.

ENVIRONMENTAL ETHICS
A Dialogue of Cultures

R.P. MISRA
Editor

CONCEPT PUBLISHING COMPANY, NEW DELHI - 110059
in Cooperation with
SUSTAINABLE DEVELOPMENT FOUNDATION
Delhi - 110 007

ISBN 81-7022-570-1

First Published 1995

Published and Printed by
Ashok Kumar Mittal
Concept Publishing Company
A/15-16, Commercial Block, Mohan Garden
NEW DELHI - 110059 (India)

PREFACE

The Sustainable Development Foundation (SDF), New Delhi, organized a pre-UNCED Consulative National Symposium on Environmental Ethics on 1st March, 1992. The Symposium was co-sponsored by the University of Delhi, Commonwealth Human Ecology Council, London, and Gandhi Smriti and Darshan Samiti, New Delhi. It was assisted by the Ministry of Environment and Forests, Government of India; UNESCO-UNEP International Environmental Education Programme (IEEP), Paris; University Grants Commission, New Delhi; and Indian Council for Social Science Research, New Delhi.

This volume contains the gist of discussions, deliberations and comments of the participants. It also includes the addresses by eminent persons and the recommendations of the symposium. It was earlier brought out by the SDF for limited circulation. However, the demand for the book has been so overwhelming that it became necessary to bring out a new edition with appropriate changes and improvements. A comprehensive Index has been added for the convenience of the reader.

The papers contributed to the National Symposium on Environmental Ethics have now been published under the title 'Environment Development and Education; Cultural, Ethical and Spiritural Perspectives'. Those interested in the volume may write to the editor.

I am confident that the volume will make a significant contribution to the on going debate on ecology, environment and sustainable development.

I am thankful to Shri Ashok Mittal for taking up the publication of the new edition.

R.P. MISRA
President
Sustainable Development Foundation.

FOREWORD

UNESCO considers environmental ethics as an integral component of the ultimate goal of environmental education (EE). Thus, the objective of the "Symposium on Environmental Ethics" reported in this volume coincides with the goals fostered by the UNESCO-UNEP International Environmental Education Programme (IEEP). Several studies to this effect have been prepared and an international consulation meeting was organized in cooperation with the Egyptian National Commission for UNESCO in March, 1991.

I would like to thank Prof. R P Misra for inviting UNESCO to collaborate in the preparation and organization of this "Symposium on Environmental Ethics" and am pleased that UNESCO was able to make a contribution to the two volumes, and thus further the debate on this timely subject.

The agenda and corresponding papers of the Symposium have covered a wide range of relevants problems, and issues in the context of environmental ethics. The philosophical foundation of environmental ethics traces the development of philosophy and science and considers scientific knowledge as physical power which should be used with ethical responsibilities. In this context, environmental ethics is considered to be a code of behaviour which ensures human progress without jeopardizing ecological balance. Humans are an integral part of the environment and we must learn to live in harmony with nature and our fellow beings. To be able to live together in peace is the basic "ethics of all ethics". This is the foundation and goal for environmental ethics.

Other important topics covered are: Environmental Ethics and Emerging World Order; Laws and Environmental Ethics; Tribal and Rural Experiential Knowledge and Environmental Ethics; Culture, Religion and Environmental Ethics, Population Growth and Environmental Ethics; Science, Technology and Environmental Ethics; Education and Environmental Ethics; and Development Styles and Environmental Ethics. The richness of the deliberations on the above issues signals the deep knowledge, extensive experience and, more important, the commitments of a group of educators to environmental ethics and its development as the basic ethics of all ethics for humankind to live in peace with each other and with the environment.

It is certain that the development of environmental ethics takes time and its starting point should be as early in life as possible. This signals the importance of pre-school education and the role of parents at home to live environmental ethics for effective learning and practice by children. Formal education, with emphasis on environmental education and its practice, should be built upon pre-school experiences. Mass-media should promote environmental ethics through various accurate messages. Cultural and religious institutions may play an important role in the further development and practice of environmental ethics because its seeds are found in all of them.

The proceedings of the Symposium have been well reflected in the Executive Summary. It reflects the major concepts covered and recommendations produced at the Symposium. The proceedings, especially its executive summary, will be of use to politicians, administrators, development experts and educators for the understanding of environmental ethics, its place and need in contemporary society and mechanisms for its development by environmental education for all types and levels of education. These deliberations on various aspects of environmental ethics could serve as mirrors for readers for checking, correcting and developing on their part the needed ethical values and responsibilities in everyday life and in the decision-making process with respect to the holistic nature of the environment, sound management and rational use of resources and protecting and improving the environment for a decent, sustainable and equitable quality of life of the present and future generations.

COLIN N. POWER
Assistant Director-General
for Education

ACKNOWLEDGMENT

I am grateful to the United Nations Educational, Scientific Cultural Organization for the intellectual and financial support in making this publication possible. I am particularly thankful to Mr. Kamal Nath, Honourable Minister for Environment and Forests, Government of India, Mrs. Zena Daysh, Executive Vice-Chairman, Commonwealth Human Ecology Council, London, Dr. A. Ghafoor Ghaznawi, Chief, Environmental Education Unit, UNESCO, Paris, and Prof N. Radhakrishnan, Director, Gandhi Smriti and Darshan Samiti, New Delhi, for personal involvement, support and sponsorship. Prof Ram Reddy, Chairman, University Grants Commission, New Delhi, and Prof D.N. Dhanagare, Member Secretary, Indian Council for Social Science Research, New Delhi, in spite of the financial constraints, supported the Symposium. We are grateful to them.

R.P. MISRA
President
Sustainable Development Foundation

CONTENTS

OBSERVATIONS AND RECOMMENDATIONS

Executive Summary

1. Realizing the seriousness of ecological problems of the world today, the symposium considers it imperative to evolve an environment sensitive value system. As a first step in this direction there is need for a cross-cultural dialogue to distill from each culture all those values which promote empathy between humans and nature and humans and humans.
2. All ancient cultures and civilizations are founded on creative human-nature relations with a built-in mechanism to correct minor ecological disturbances. Rapid industrialization, urbanization and population growth have given rise to imbalances which cannot be corrected through self-correcting processes. If these three forces are not adjusted to conform to the laws of ecology, environmental pollution and economic poverty cannot be avoided.
3. The symposium does not see any chance of correcting the current ecological imbalances and consequent human suffering through new technologies alone. Unless people seek meaning in life and living beyond the narrow ends of consumerism and exploitation of nature and fellow human beings, new technology will bring only new and more miseries. The symposium, therefore, recommends a more austere style of life laying greater emphasis on cultural enrichment and less on the amount of goods consumed. Such a life-style requires a new system of ethics where in the concern for life in whatever form it exists in nature is the uppermost.
4. The environmental laws can do a lot to improve the deteriorating ecological situation. The laws can, however, be obeyed only if they are in tune with the value system of the people. As such while more formal laws should be enacted to stop the rape of the earth and exploitation of the poor, the need to evolve, codify, and disseminate an ethical code of conduct based on a judicious integration of the rights and duties cannot be overemphasized. Laws must support ethics, and ethics the laws. The symposium

recommends more research on the junction and disjunction between laws and ethics.

5. Environmental ethics cannot be imposed by law. It has to be articulated, systematized, codified and brought to the doorsteps of each and every individual. This needs a comprehensive environment sensitive educational policy and curricula. Environmental education has to begin from the family. Other channels are:

   Formal educational processes (primary to university)
   Informal cultural institutions
   Communication media

   The symposium recommends that all these channels throughout the world should be used to educate the people about the environmental sensitive values. UNESCO has played a catalytic and very vital role in the past in the field of education. It must continue to give the lead in this direction too.

6. The current development models are counter-productive. They are insensitive to ecological and human needs. In fact, it is the blind pursuit of these models which has given rise to pollution and poverty in the Third World countries and led to the destruction of nature and the emergence of consumerist culture in the developed countries. The symposium recommends that search for a new style of development oriented not only to economic growth but also to cultural enrichment must be intensified. And in this search, the potentials of the Gandhian holistic approach and the Principles of Human Ecology can be used as guiding principles.

7. Population pressure on the natural resources is a serious problem faced by the Third World countries. The current approach to population control does not augur well. Population growth and human development are negatively related. And population growth and environmental degradation are positively linked. Therefore, improvement in the quality of life alone can stem the rapid population growth and environmental destruction. The Symposium, therefore, recommends that population planning, human development planning and environmental planning must be enmeshed into a single planning process.

8. Any attempt by the developed countries to link aid and trade with

environmental policy will be counter-productive and harmful to the Third World countries. This will amount to imposition of new conditionalities in international relations. There must be an international code of environmental ethics which each country should follow. Such a code must, however, be sensitive to the inequities in the world today and must not come in the way of development of the Third World countries.

9. Most of the current environmental problems—green house effect, ozone layer depletion, emission of toxic gases like CFC, deforestation, desertification, energy crisis, etc., have been created by the industrialized countries. The poor countries have benefited the least by treading in their footsteps but the consequences of the ecological disasters are being borne by them equally.

10. Globalization of the economy and ecology is a step in the right direction. But it should not scuttle the future development of poor nations. National sovereignty over natural resources can be given up only if there is a fair and respectable equity in development the world over.

11. Science and technology cannot be blamed for all the ills of the society. Much depends on how we use them. And how we use them depends on our value system, which in turn determines the ethical code of behaviour. The Symposium recommends that instead of trying to look for scapegoats in science and technology, we should introspect and work to change our values. Thus alone can a pollution-and-poverty-free society be ensured and the ill-effects of science and technology removed.

12. The Symposium recommends that voluntary organizations all over the world should be more directly involved in the search for practical and equitable approaches to the resolution of development dilemmas using the ethical codes of different communities, cultures, religions, etc. They must involve the poor, the women and the youth to fight for an ecologically sound development process and environmentally friendly value system.

13. The Commonwealth of Nations is the most representative organization integrating the interests of the rich and poor nations of the world. Its potential must be fully utilized to foster global togetherness on issues relating to environment and development. The Commonwealth Human Ecology Council and its network can be effectively utilized for this purpose.

14. UNCED should pay more attention to the poverty-pollution nexus in the Third World countries and should not be used as a forum to slow down their development in the name of environmental protection. The new world order must help these countries improve their economy and society; it must also encourage the developed countries to come down to a reasonable level of consumption and production eliminating waste of natural resources and reducing environmental pollution. Developed countries must provide resources and technology to the Third World countries so that they progress and eliminate poverty using non-polluting technologies.
15. There is a need to integrate science and spirituality. Neither de-spiritualized science, nor unscientific spirituality is the answer to the problems of the future. How to do it, is the crux of the problem. This Symposium raised these issues, and discussed them. More questions have now emerged. Let the search for answers be on.

> *Evidence now supports the vision of the poet and the philosopher that plants are living, breathing, communicating creatures. It is only we, in our blindness, who have insisted on considering them automats.*
>
> —From The Secrets of Life of Plants
> by Peter Tompkins and C. Bird.

1

INAUGURAL ADDRESSES

1

INAUGURAL ADDRESSES

Welcome Address

Prof R P Misra
Director, Gandhi Bhavan, Delhi University, and President, Sustainable Development Foundation, New Delhi.

It is a unique privilege for me, ladies and gentlemen, to welcome you to the National Symposium of Environmental Ethics. It is a pleasure for Gandhi Bhavan, University of Delhi, and Sustainable Development Foundation to have you here.

That the Symposium is being attended by a good number of decision-makers and intellectuals like Shri Kamal Nath, Honourable Minister for Environment and Forests, Shri K C Pant, former Minister for Human Resource Development, Shri B N Pande, former Governor of Orissa, Prof S K Aggarwal, Secretary, Association of Indian Universities, Prof Upendra Baxi, Vice-Chancellor, Delhi University, Prof Vidya Niwas Misra, Vice- Chancellor, Kashi Vidyapeeth, Varanasi, Prof B M Shukla, former Vice- Chancellor, Gorakhpur University, Prof D N Chaturvedi, former Vice-Chancellor, Kashi Vidyapeeth, Prof K D Gangrade, former Pro-Vice-Chancellor, Delhi University, Prof T K N Unnithan, former Vice-Chancellor, Rajasthan University, Jaipur, Dr B R Nanda, Prof B K Roy Burman, Prof R R Singh, Prof Zafar Raza, Father Anto, Prof Rukmini, Prof Indra Pal, Prof H N Misra, Prof S K Sinha, Shri Anil Agrawal, and Prof Ashish Bose, all of them very senior social scientists and scientists; overseas colleagues like Madam Zena Daysh, Dr Ghafoor Ghaznawi, Prof Poser and a large number of environmentalists, is indicative of the importance of the theme of the Symposium. It was very kind of Shri Kamal Nath, Hon'ble Minister for Environment and Forests to have agreed to open the Symposium. Shri Kamal Nath has made the Ministry of Environment an active instrument of environmental protection. Without being populist, he has brought about a sea change in the functioning of his ministry and its impact is already being felt in all those sectors of the economy which disturb the ecological balance and pollute the environment.

Among the pioneers of the Human Ecology movement, the name of Mrs Zena Daysh, the founder and executive Vice-Chairman of the Commonwealth Human Ecology Council, stands out most prominently. Mrs Daysh, a New Zealander by birth, is now a world citizen. She is everywhere and all the time spreading the message of love and Ahimsa and appealing for due respect for all that exists in nature.

Mrs Zena Dayash began as a physiotherapist. The instinctive love for humanity and the desire to restore individuals to health soon turned into a movement to restore the loving bond between humans and nature.

The foundation of this movement was laid in 1952 with the establishment of the Committee on Nutrition in the Commonwealth Secretariat. Distinguished persons like F E Schumacher, Sir Julian Huxley, and many others joined hands under her convenership to produce results which ultimately led her to formulate what she calls the holistic development idea.

In the sixties Mrs Daysh introduced the holistic development idea to the island of Malta. The first Commonwealth Conference on Development and Human Ecology was organized in Malta. The Commonwealth Human Ecology Council (CHEC) was born there. Pioneering citizens' programmes and university courses were formulated.

Since the early seventies, Mrs Daysh has been invited to attend the Commonwealth Heads of Government meetings as a representative of non-governmental organizations. CHEC was directly identified in the 1989 CHOGM Final Communique and the Commonwealth leaders acknowledged its long established work. President Hoyte of Guyana set aside one million acres of pristine rain forest for research and action for sustainable tropical forestry. Mrs Daysh has been active in building a programme to associate the Amerindian people with such a major project.

Her ideas of life full of values of caring and sharing is an illustrious example of selfless social service worthy of emulation. Her life is a living testimony to the worthiness of following the path of Buddha and Gandhi.

Dr A Ghafoor Ghaznawi, Chief of the Environmental Education Unit, UNESCO, Paris, has been a source of encouragement in the conceptualization and organization of the Symposium. It was at his suggestion that Mrs Daysh desired that a conference be organized on Environmental Ethics before UNCED at Rio, and preferably prior to the Commonwealth Consultative Conference slated to be held at Taplow Court, London, in April, 1992. His participation brings to the Symposium a vast experience, expertize and knowledge UNESCO has acquired in the field of environmental education.

Mr B N Pande, Vice-Chairman of the Gandhi Smriti and Darshan Samiti and former Governor of Orissa, and a sitting member of Parliament, has been most helpful in the organization of the Symposium. Dr N Radhakrishnan, Director, Gandhi Smriti and Darshan Samiti, is unfortunately not attending the Symposium due to ill health. He deserves special mention. But for him, the Symposium would have not been possible.

This Symposium is not centred on environmental pollution and its abatement. It centres on the question: how to create a social order which protects the nature and the basic human rights at the same time; in other words, how to integrate the economy and ecology; material and ethical progress; and science and spirituality. It has been organized with the belief that Environmental Ethics has the potential to help create such an order.

Environmental Ethics is an internalized code of behaviour that produces an ecologically sound development process. Each culture has its own environmental code to maintain a harmonious relationship between man and nature ensuring continuing material and cultural enrichment of the society.

Industrialism and consumerism, and the value systems they promoted, commercialized both nature and human beings. Man became a factor of production, and nature a resource for exploitation. The production of farms and factories increased but at a great ecological and human cost. The earth has now become an endangered planet. In spite of ever new goods and gadgets, millions of people continue to be hungry and destitute. Science and technology have placed an enormous power in the hands of man but, without a moral code of conduct, its creative potential has remained dormant.

Sensing the dangers ahead, a search is now on for new models of development, the models which can sustain life on earth and which can ensure human progress. Environmental Ethics constitutes the bedrock of such models.

Development is not the concern of governments alone. More importantly, it is the concern of the people. A new global cooperative endeavour is on the anvil, and the people's power from the grassroots is already on the ascendance. We can see this trend all over the world. The people act, as they think. And the thinking of the people is largely determined by the values they cherish. What man will do to his environment will depend more on his values than on anything else. Thus, the urgency of examining the ethical bases on which people in different cultures/

situations interact with nature and produce various processes of construction and destruction needs no emphasis. This Symposium on Environmental Ethics provides the forum for a meaningful dialogue of cultures.

Human society is no longer compartmentalized into local communities. World has now become a global village. Yet the building blocks of this enlarged society are local communities. The environmental code which guides local communities, therefore, must constitute the foundation of any global code of Environmental Ethics. We have to learn from the communities and groups that maintained ecological balance. We have also to see how the global perspectives can help us integrate the local values into a broader code of environmental ethics.

The Symposium is, therefore, focusing on three aspects of the issue: (1) great traditions of the world and environmental ethics; (2) education processes and environmental ethics; and (3) the implications of economic development models for the earth's ecosystem on the one hand and human survival and development on the other.

As such, the environmental code of behaviour has to be seen at three spatial levels. The basic level is the community. How the local communities interact with nature and how their approaches and behaviours help use nature judiciously. Many traditional communities had evolved in the past sustainable agriculture at a low level of production. How a similar system of agriculture be developed with a higher level of production? This question needs thorough examination.

The ethical codes of these small local communities constitute the cultural concensus which is reflected in traditions, religions, cultural ethos, language, literature and the day-to-day life of a larger society. They form the cultural realms or great cultural traditions. These great traditions reinforce the values of small communities and those of the small communities enrich those of the larger cultural traditions.

This is how the world functioned until the advent of the Industrial Revolution in the West. This revolution changed the situation drastically by integrating the world of cultures into a world of colonial empires. In due course, the empire died, thanks to the inner contradictions inherent on this predatory concept. It did, however, give rise to a global society. Humanity stands united today in physical terms but it continues to be divided in human terms. Conflicts and wars continue to rage in the world. We are not in peace.

The conflict that is inherent in man's quest for happiness through material advancement has taken a variety of forms. There are world wars, there are religious wars, there are social conflicts, there are psychological tensions. The whole human society is getting torn into pieces. But above all these man has waged an incessant war against nature endangering the very survival of life on the earth.

We name witnessed two World Wars in this century and experienced the devastation through which many nations of the world passed. But the ill-effects of these wars were not fortunately so lasting. Countries like Germany and Japan which were virtually destroyed were rebuilt within a decade or so. But the silent war against nature that man started in the 18th century has only now acquired serious proportions. Man survived the fury of the two World Wars; but he would not survive the fury of the war he has waged against nature. The approaching ecodisaster is visible even to those whose vision is imperfect. Therefore, before it is too late, man must return to sanity. The other name of this sanity is the conduct of life in consonance with the laws of ecology.

But it would not be enough to have only ethical values to ensure harmonious relationship between man and nature at the community level; we have to have a code of conduct for this purpose at the national level and also a global code of environmental ethics to guide the behaviour of the global society.

All these aspects demand earnest attention and honest action at all levels: community, group, society and humanity. They also raise a question: How to convert the human responces to emerging environmental challenges into new values so that humans and other living species can endure on earth and live in harmony and hope, freedom and dignity? The symposium is a pre-UNCED 1992 national conference. Its conclusions will constitute an input to the deliberations at Rio, Brazil.

Ladies and gentlemen, I have great pleasure in once again welcoming you all to the Symposium.

Keynote Address

Mrs Zena Daysh
Executive Vice-Chairman, CHEC, London

The Commonwealth Human Ecology Conference held in Jaipur in 1991 threw up a number of ideas for follow-up action. One of them was to hold a conference on Environmental Ethics, prior to the United Nations Conference on Environment and Development (UNCED) to be held at Rio de Janeiro in June 1992, and preferably before the Commonwealth Consultative Meeting to be held in London in April-May, 1992. So I am really pleased that Professor Misra took the initiative and organized this national meet which has turned out to be international as there are a number of overseas participants also.

This conference is organized to discuss some vital issues involving human survival and development. We should always extend a helping hand to all such endeavours. So I am here today amongst you. Coming to India has always been a pleasure to me, for it gives me an opportunity to learn more about this country's concept of oneness, unity and holistic world view. It also gives me a chance to renew my contacts in India.

India has always fascinated me. This country gives a strong message of compassion. When I met the Prime Minister of India, Mr P V Narasimha Rao, at Harare during the last meeting of the Commonwealth Heads of State, I inquired of him of the environmental situation in India. He said he had a very strong Minister for Environment to take care of it. I am glad that Mr Kamal Nath, Minister for Environment, is here today to open this conference.

Dr Ghafoor Ghaznawi has consistently helped us to put our ideas into action through the support and sponsorship of UNESCO. He has a passion for and a mission of Environmental Education. I am glad that he could find time to come here and join us in the deliberations. Friends, I was born in New Zealand and finally landed in England and had the opportunity to work for Human Ecology. For me, India has a lot to offer to the world. India gives a strong humanistic message and I am confident that it will continue to inspire the world with its message of peace, harmony and unity.

The Commonwealth of Nations comprises more than one quarter of the world's population and we find it an ideal global working model for consensus among disparate cultures and nations. There is no continent, no culture, no religion and no ethnic group which is not representated in the Commonwealth of Nations. And, therefore, this institution has great responsibilities. The Commonwealth's unique responsibility towards ecological balance and human welfare was clearly recognized and given practical expression at the Commonwealth Heads of Government Meeting (CHOGM) in Kuala Lumpur in 1989, when President Hoyte of Guyana offered one million acres of pristine tropical forests for special status and study.

Central to the phenomenal changes, which are currently taking place in the world, are the voices of the ordinary people. Therefore, we stress the need to gain the full support of the power and energy of the spirit of the people both locally and globally through communication, education, and consultation, as a pre-eminent contribution to economic, scientific and technological considerations towards sustainable development.

The Commonwealth Human Ecology Council (CHEC) has thirty years' experience of applying human ecology to environmental challenges. CHEC organizes and sponsors urban and rural projects in these fields, focusing on skills, training and self-reliance. The cornerstone of these projects is the use of local expertise and indigenous knowledge. CHEC received endorsement for its activities in the final communique of the 1989 Commonwealth Heads of Government Meeting. CHEC responded by organizing regional seminars during 1991, Caribbean seminar at Georgetown, Guyana in February, 1991, Asian seminar in Jaipur, India, in May, 1991, and a synthesizing seminar at Vancouver, Canada, in July, 1991. CHEC is providing the structure and intellectual environment for scientists involved in the Commonwealth sustainable tropical forestry programme in Guyana, to work with the Foundation for Ethnobiology and gain the full support of the indigenous Amerindian people.

It has been a consistent endeavour of CHEC to promote a dialogue of cultures to bring about a consensus on how to bring about peace between nature and man, and man and man. Human ecology relies on love that pervades each and every individual to strengthen the feeling of oneness, which alone can ensure the survival and progress of mankind.

I am really grateful to Gandhi Bhavan, Delhi University, Gandhi Smriti and Darshan Samiti, and Sustainable Development Foundation for organizing this pre-UNCED Meeting. The conclusions of this symposium will form the background material for the pre-UNCED Commonwealth Consultative Meeting to be held at Taplow Court, England, next month. This symposium gives us an opportunity to evolve a global code of conduct for the member countries of the United Nations to ensure ecological balance and environmental quality and to strengthen the community and cultural values which protect the nature while at the same time ensure a dignified and materially as well culturally enriched human life.

Keynote Address

Dr A Ghafoor Ghaznawi
Chief, Environmental Education Unit, UNESCO, Paris

I am thankful to Prof Misra for inviting me to participate in this very important consultation meeting. It is a pleasure to be here in India which has done so much to arouse the conscience of the world against environmental destruction. I convey the greetings of the Director General of UNESCO for the success of the conference.

This Symposium is focusing on Environmental Ethics which makes an extraordinary contribution to contemporary societies, to ecology, and to environment. Ethics constitutes the cornerstone of civilizations. Any attempt to transform our own civilization and to ensure peace and harmony cannot ignore the ethical issues. I am glad that Prof Misra decided to organize this Symposium on this theme.

A year ago, I had the opportunity of meeting Dr Upendra Baxi, Vice-Chancellor of Delhi University. We discussed the United Nations Conference on Environment and Development to be held in Rio de Janeiro in June next. Our discussion centred on Environmental Law, Environmental Education, Women's Education and Environmental Ethics. Subsequently, the Jaipur seminar concluded, among other things, that ethical dimensions of environmental issues should get priority attention. Then there was a Consultation meeting on Environmental Ethics at Cairo in March 1991. It was organized by UNESCO. These meetings further strengthened the UNEP-UNESCO Environmental Education Programme with respect to environmental ethics. This Symposium is a follow-up of these activities. It is also designed to move further in strengthening the Environmental Education Programme. The emphasis, as I see the programme, is on how to inculcate environmental ethics in the people.

Environmental ethics is the ultimate goal of environmental education. Environmental education fosters awareness, ecological knowledge, attitudes, values, commitment, skills and environmental responsibilities in individuals and nations for a sound use of the earth's resources and for protecting and improving the environment for the present and future generations. The Indian culture is known for sound and widely accepted tenets of environmental ethics.

Lastly, I would like to underline the important statement made by Prof Misra in his Welcome Address. Humanity and environment are not two discrete phenomena. They constitute an integrated life-cycle system and hence have implications for each other. Environmental education is the means to strengthen the bonds between the two.

Environment is the life support system of humanity. Clean, healthy and safe environment is a human right. Environmental ethics is a code of behaviour to protect this right.

Inaugural Address

Shri Kamal Nath
Honourable Minister of Environment and Forests, India

I am very happy to be asked to inaugurate this Symposium on Environmental Ethics and am grateful to Prof Misra for his kind invitation. I would also like to congratulate the Sustainable Development Foundation, Gandhi Smriti and Darshan Samiti, Gandhi Bhavan, Delhi University, and Commonwealth Human Ecology Council for their initiative in organizing this very important discussion. Environment is a subject of frequent discussion but the emphasis is usually on environmental problems and their management with the focus on the problems of pollution and pollution control. It is heartening that today's symposium looks at these issues from a different viewpoint.

As I understand it, environmentally ethical conduct is based on ecologically sound principles which lead to environmentally sensible decisions. In every society, the people and institutions created by them need to appreciate the requirements of the future as well as the requirements of the present. They will need to understand the intergenerational consequences of any decision involving the use of natural resources. In this lies the essence of environmentally ethical behaviour.

The needs of economic development are often posed as conflicting with the needs of environment. This need not necessarily be so. Over the centuries, some human societies have reached an environmental equilibrium in which it was possible to draw upon nature without destroying its resource base. With increase in population and with the availability of technologies that could exploit earth's resources in bulk, that equilibrium is now lost. We now find an intolerable pressure of population on the environment. In our country, as in large parts of the world, there are still terrible problems of poverty and deprivation. Most solutions to these problems involve the use of what we have by way of natural resources. However, it is very clear that we cannot use these resources without paying heed to their sustainability. These are very difficult decisions. We need very sound ethical principles to arrive at right decisions.

Twenty years back in 1972 at the Stockholm conference on environment, Mrs Indira Gandhi said and perhaps for the first time that the problem of poverty is one of the principal problems of environment. Twenty years later, in June this year, we go to the Earth Summit in Rio de Janeiro. What has happened in this transit of twenty years? Earlier where thousands of kilograms of chemicals were needed to destroy the environment, today only a few kilos are needed to inflict far greater damages.

In her inaugural address to the Non-Aligned Conference in 1983, Mrs Gandhi said: "Some people still consider concern for the environment an expensive and perhaps unnecessary luxury. But the preservation of the environment is an economic consideration since it is closely related to the depletion, restoration and increase of resources. In any policy decision and its implementation, we must balance present gains with likely damage in the not too distant future. Human ecology needs a more comprehensive approach." This is a message we cannot forget. I, therefore, fully agree with Mrs Zena Daysh that Human Ecology must go beyond the tunnel visions of technical disciplines and look at the environment from a broad human perspective.

Every society has its own code of conduct, its own norms and its own concept of desirable behaviour. Our ancient scriptures and poems, the rituals and celebrations that still form a crucial part of our life, particularly the tribal life system, reflect these values. I am happy to note that today's symposium is to discuss the importance of rural and tribal knowledge and ecological values, and also the environmental codes enshrined in different religions and cultural practices.

Any initiative that we take today in the field of environmental education or awareness generation must necessarily come from our own cultural heritage. Countries in the developing world are notoriously susceptible to the impact cultural practices from the industrialized world. While there is great deal to be learnt from the intellectual and technological advances made by the countries of the West, we must beware against importing the environmentally destructive life-style that prevails in those countries.

The question of how well we are able to use our natural resource base and how wisely we make decisions are not questions to be answered by governments alone. We know that any development effort requires the participation of people. People are guided not only by the laws we create, but also the codes of conduct and beliefs of society. We cannot expect laws to be followed merely because we have legislated them. It is also necessary for the people to believe them to be right. This is where the process of environment education becomes important.

Most of you are familiar with the global issues in the environment sector such as climate change, depletion of the ozone layer, the loss of biodiversity, etc. The need for environmentally ethical conduct by the countries of the world in response to these issues is also very important. These are issues that we hope will be discussed at the UNCED deliberations in Rio later this year. In today's symposium, as in Rio, we will be debating these fundamental questions. The nature and the very existence of a viable future depends on the answers we have to these questions.

I would like to put forth before this august gathering a few points for discussion and debate. The *first* relates to the incompatibility of the Western capitalist system with the idea of conservation. And the *second* relates to the obvious need of prescribing a way of life based on simplicity as the panacea for all our ecological and environmental problems.

Here I would like to relate to you a very interesting anecdote. A Western environmentalist was recently in Delhi to lecture us on whether the life-style of the advanced Western nations would be an issue during the UNCED. The two points which I have just mentioned were put to him. He vehementally agreed with their validity. But this dilemma came to the open when he was asked whether, in his country, or in other Western nations, any serious effort had been made in schools to teach children the need for simplicity in life. He said, "unfortunately not". And he added that this was an ethical question and since his country was secular (meaning non-religious or irreligious unlike the Indian definition of equal respect for all religions) apart from some lone voices, nobody took the need for simplicity as a crucial debating point, while discussing environment.

Here I would like to point out that after a discussion of the man-induced environmental degradation, an Indian audience instinctively thinks of simple life-styles as the alternative, and the only remedy to the environmentally degrading life-style. But to a Western audience this is not a solution, not only because they presently enjoy a higher material standard of living and are loath to give it up, but also because culturally and philosophically the thought of a simple life-style is alien to them. For them this alternative does not exist; it never existed before.

This is the reason why man-induced environmental damage is interpreted by Western nations as merely a scientific and economic issue and not as an ethical or philosophical one. To us Indians, and to much of the Third World, it is only an ethical issue. The real question at the present juncture is whether environment should be treated as a pure scientific-cum-economic subject or as a scientific-cum-economic-cum-ethical issue.

On the ethical plain, historically, two streams of thinking have existed. One is the oriental thinking of human beings living in harmony with nature, and the acceptance of the need for the restraining of human desires rather than incessant attempts to satisfy them. To a Hindu, the flora and fauna, and even inanimate objects were part of the same Supreme Being of which he himself was a part. The thought of conquering nature never occurred to him. Remember Pandit Nehru objecting to the use of the word "conquest", to describe the scaling of Mt Everest. Buddhism, Confucianism, Islam and Christianity celebrated the concept of living in harmony with nature. But, in contrast, the Western philosophy revelled in the idea of conquering nature.

As I understand it, a social philosophy that ignores the basic laws of the science of ecology cannot be said to be either adequate or relevant. Such a philosophy falsely promises to guarantee an infinite, inexhaustible supply of material goods. Clearly, the criteria of "good society" and "good life"—as an ever expanding cornucopia of economic growth, possession and consumption of external material things—will not do. This is because the goal of growth and consumption, forever more, is both a physical impossibility and a conceptually incoherent ideal. We need to construct a philosophy and emanating from it a code of ethics which keeps in view the principles of ecology and finiteness of our resources.

I do hope that the Symposium will make us more greatly alive to the question of environmental ethics. Environmental issues are being discussed in a variety of for a, but this is the first time that I am attending a meeting on environmental ethics. I am confident that its outcome will help us move a step forward in resolving the environmental problems we face today.

Presidential Address

Shri B N Pande
Vice-Chairman, Gandhi Smriti and Darshan Samiti, New Delhi

Shrimati Indira Gandhi had an intense love for and a total commitment to the protection and preservation of the natural environment around her, her countrymen and people of the world. Whether it was the riotous combination of colours in foliage and flowers or whispers of a perennial spring, or the humming of some of the rare species of birds in a natural habitat, or the prowl of the denizens of the forest, she had an instinctive attunement to and admiration for the whole of nature.

She was quick to realize the dangers of pollution and degradation of environment in a world of increasing urbanization and industrialization. She set up a National Environment Committee to act as a nodal point in the Government of India for consideration of environmental issues. It is on account of her intense love for protection of some of the rare species of wildlife which are fast getting extinct that the Wild Life Protection Act was passed by both Houses of Parliament on September 9, 1972.

In 1980, she took the initiative to set up a full-fledged Department of Environment at the Centre under the direct charge of the Prime Minister. This was also the year which saw the birth of the Conservation of Forest Act, 1980, which in many respects was a welcome departure from the traditional and regulatory approach of the Indian Forest Act of 1927. Followed in quick succession a chain of efforts, starting with the creation of new Departments of Science, Technology and Environment by the State Governments, creation of a new Department of Forests by delinking it from the Department of Agriculture at the Central Government level, enactment of new laws to stringently deal with the problems of air and water pollution, setting up of boards for more stringent enforcement of the law, strict penalties for destruction of wildlife, new laws and institutions for town planning with a futuristic orientation, better drainage and sewerage for a cleaner city life, intensive measures for decongestion and beautification of cities through removal of slums and rehabilitation of slumdwellers, so on and so forth.

None can dispute that land constitutes the most important productive asset in our country as elsewhere in the world. It provides an immediate access to production apart from conferring a position or status on those who own land. Management of this most important productive asset must, therefore, be viewed in its totality. In other words, it is not enough to think of increasing productivity of agricultural land merely through improved irrigation and application of larger dosages of fertilizers and pesticides. It is necessary to simultaneously plan and implement measures for arresting erosion and denudation of top soil through intensive soil conservation measures such as contourbunding, check dams, water-harvesting structures, afforestation, etc.

Today, while around 47 million hectares of our total agricultural area of 143 million hectares are classified as irrigated, the remaining area is vulnerable to recurring floods, droughts, low productivity and resultant poverty. This coupled with the denudation of forests at the rate of 1.5 million hectares per annum and causing disappearance of vegetable cover almost from one-third of the country's total land area (converting in the process the productive green covers into deserts) presents a very dismal scenario. We are, therefore, confronted with the twin problems of substantially increasing food production form the present level of around 1.7 ton per hectare to at least 4 tons per hectare through the application of a package of scientific land management practices on the one hand and preventing recurrence of floods and drought situations through afforestation, watersheds and increasing the vegetal cover by preventing illicit and illegal felling of trees, on the other.

Next in order of importance to land comes water which holds the key to life. Our young and far-sighted Prime Minister, late Shri Rajiv Gandhi, in his very first broadcast to the nation hinted at an organized effort to purge the Ganges of the massive pollution—which is taking place today due to an unchecked discharge of industrial effluents, human and animal wastes—and creation of a Central Ganga Authority. This was a good beginning in the right direction of arresting pollution of water of one of the most sacred rivers of India. The Central Ganga Authority has set in motion organized efforts on the part of public health engineering and civic authorities of 27 riverside towns to grapple with a formidable problem of sewerage treatment and pollution control which had been eluding us for a long time. The Ganga, however, is only one of the many perennial streams which continue to be polluted by effluents and wastes and, therefore, the problem of polluion of all these streams deserves an equally urgent and important attention.

Having dealt extensively with the management of land and water resources, I would like to turn my attention to the extremely adverse effects of industrialization specifically by way of air and water pollution and how to control these hazards. The Central Water (Prevention and Control of Pollution) Acts were passed by both the Houses of Parliament in 1974 and 1981 respectively. These were enacted with the laudable objective of preventing and arresting the incidence of air and water pollution and have been adopted by most of the States including ours. State (Prevention and Control of Polluion) Boards have been set up by most of the State Governments to give effect to these laudable objectives. They are intended to ensure that utilization of natural resources for material progress and advancement does not result in a decrease in the quality of human life. Despite this, however, there are certain risks and hazards inherent in the manufacturing process of number of industrial units which, if not kept under control, will unfold themselves, causing untold damage and misery to the life of industrial workers as well as of the common man. What happened to the Union Carbide plant at Bhopal on December 2, 1984, is only a reminder of the grim reality lying ahead. This man-made disaster left more than 2,000 dead, thousands injured and maimed for life, apart from wiping out human settlements from the surface of the earth. Several committees and commissions have been set up to pin point the factors which led to the tragedy and volumes have been written since then but the cup of misery and suffering of the people affected by the calamity remains full.

Despite the enactment of series of legislations, existence of enforcement agencies like the Inspectorate of Factories, and creation of institutionalized bodies like Boards for Prevention and Control of Pollution, the incidence of pollution and hazards thereof continues unabated.

Whether it is possible to achieve the goals of development without uprooting the local people from their historic-religious-cultural-economic milieu is the enigma before every environmental planner and development administrator. This enigma gets more pronounced as we turn our attention to other areas of environmental planning such as modern chemical-based farming which robs soil of its vitality; exploitation of virgin forests for industrial purpose which can hardly be replaced; exploitation of marine resources for augmenting the consumption of animal protein; unplanned and unregulated urban centres which devour the soil for bricks, sand and metal, and which the present the agonizing spectacle of slum life.

I quote below a few stanzas from *Bhumi Suktam* of the *Atharva Veda* which was composed thousands of years before but is as fresh and life-giving as then:

> Earth in which are the sea, the river and other waters,
> In which food corn-fields have come to be,
> In which live all that breathe and that move,
> May she confer on us the finest of her yield,
> All sustaining, treasure-bearing, firm-staying place,
> Gold-breasted, home of all moving life...
> Pleasant be thy hills, O Earth,
> Thy snow-clad mountains and thy woods,
> Earth, in which the waters, common to all,
> Moving on all sides flow unfailingly, night and day,
> May she pour on us milk in many streams and endear us with lustre.
> We invoke all supporting Earth
> On which trees, lords of forests, stand ever firm,
> Rock, soil, stone and dust,
> Earth is held together and bound firm,
> To her my obeisance,
> To the gold-breasted earth,
> May that have quick growth again,
> O purified, may we not injure thy vitals or thy heart.

The concept of Earth as the sustainer of the human race, of the race itself as one family and of the necessity to nourish and protect the earth from despoilation is thus deeply rooted in our ancient thought and cultural heritage. That human beings as the children of God, as the finest and the best in God's creation hold the key, through contentment and self-abnegation (in sharp contrast to acquisition through avarice and greed) to the sustenance of the universe has been forcefully brought out in these words of the *Isavasyopanishad*:

> Please remember that this world is the Lord's creation,
> And everything that exists here is His,
> He pervades the whole universe.
> Be content with whatever is left by Him
> And do not greed the wealth of others."

These words beckon us to a realization of the sad truth that man's greed has created the environmental problem and also the problem of pollution. May this Symposium provide that awakening, so that mankind learns to replace acquisitiveness and greed by self-abnegation, lust by restraint and self-discipline, destruction by regeneration.

2

PHILOSOPHICAL FOUNDATIONS OF ENVIRONMENTAL ETHICS

2

PHILOSOPHICAL FOUNDATIONS OF ENVIRONMENTAL ETHICS

Opening Address

R P Misra

Philosophy is organic to the nature of man. To ponder, to reflect, and to reason out is as much a part of his nature as to love, hate and strive.

Some of the questions that philosophy tries to answer are :

1. Is this universe a random collection of atoms or it is designed?
2. Is the evolution of life a mere chance or it is purposive?
3. Is there a soul?
4. Is there God?

Philosophy defends reason and affirms values. It lays bare the influence of ideas. It is ideas which shape the character of the world. Things first happen in the minds of men and then are translated into action. Reasoning produces ideas and values give them direction.

Philosophy is the fountainhead of ideas. It supplies the bases for synthesizing knowledge, helps individuals to reflect, judge, and think; refines humanistic values and establishes their universality; promotes freedom, tolerance and deeper understanding of man; and helps individuals to form ideas on all problems and to find ways and means to solve them.

The Indian philosophical system is marked by a two-fold unit of outlook—spiritual unity and moral unity. Intellectual study and reasoning to know the Self must be accompanied by moral discipline. To awaken the spiritual in man and to help him realize it and thus to humanize man, is the supreme objective of all institutions.

The roots of Western philosophy lie in the Greek thought which was not far away from the contemporary Indian philosophical thinking. (5th to Ist century BC). Plato's conceptualization of Soul, physical world, God, etc, continued to influence Western thinking until about 17th century AD. Rene Descartes (1586-1650) promoted the concept of dualism of mind and matter as separate substances but accepted the omnipotence of God. Cartesian dualism ultimately gave rise to two streams of knowledge. The philosopher studied the mind, and the scientist the matter.

The traditional Christian belief in the omnipotent God and all speculations and investigations related thereto such as soul, morality, spirituality, etc., became the concerns of the Church which in due course was delinked from the so-called secular aspects of life. Scientific studies, on the other hand, laid emphasis on tangible evidence. Rigid deductive system practically superceded the inductive system. By the middle of the 19th century, the unified theory of knowledge was thrown over board and most of the physical sciences were no longer the subject matters of natural philosophy. Philosophy was left with only the nonsensible residue, i.e. epistemology, metaphysics and ethics. This approach ultimately separated social disciplines too from philosophy. They became social sciences.

Science is one of the greatest achievements of man, the noblest activity that made it possible to lessen human drudgery and unravel the mysteries of nature (in its broadest sense). In the words of Bertrand Russell, "one hundred and fifty years of science have proved more explosive than five thousand years of pre-scientific culture". It has added a fourth dimension to what is desirable: Truth, Beauty, Goodness, and *Utility*.

The spectacular achievements of science and its applied derivative technology led us to accept its omnipotence. The necessary social wisdom and enlightened conscience of man to use these powerful ideas and techniques did not develop. In fact, the pre-Cartesian man attempted to balance the knowledge of physics with the knowledge of Self. Man fought against nature but very humbly. Man and nature co-existed, each changing but enriching the other slowly and steadily.

The modern science opened new vistas. It improved and increased the powers of man several times. Proud of his achievements, man abandoned the old balance he had struck with nature and made the mighty, powerful, enduring and stubborn nature to submit to his will. He never expected this windfall; nor could he prepare himself to use the potentials of science and technology to his best advantage. The apparent victory of man over nature is now proving to be his real defeat, may be annihilation. Man has now learnt that the gifts of science and technology are not unqualifiedly good in themselves. There are so many undesirable elements wrapped up in what he thought was the unexpected gift of nature.

Every addition to power that science and technology has put in the hands of man equally involves the possibility of its misuse. He is like the new rich who has come into a fortune but is too uneducated to spend it intelligently. Knowledge is power. Scientific knowledge is physical power. Science, however, does not tell us how to use this power. It is the philosophy which talks about the Ends—what is Good and what is Evil. Science sees the physical phenomenon in bits even though in depth. Philosophy sees each phenomenon as a part of the whole that includes both Matter and Mind.

Environmental problems which we face today are largely the byproducts of our scientific achievements. Forests have been felled; animals have been killed; less privileged and weak have been exploited; soil, water, air and foods have been polluted; human body has been toxicized. We have more of more things, but less and less of those which make us happy and contented.

Science has placed at our disposal a great power. But, we must equip ourselves with such knowledge as is necessary for a proper use of it. Plato and the *Upanishads* declare that the knowledge of Good and Evil is absolutely necessary for a proper use of scientific power. "It is not the life of knowledge, not even if it included all the sciences, that creates happiness and well-being, but a single branch of knowledge— the science of Good and Evil. If you exclude this from other branches, medicine will be equally able to give us health, shoemaking shoes, and weaving clothes, seamanship will still save life at sea and strategy win battles. But without the knowledge of Good and Evil, the use and excellence of these sciences will be found to have failed us", says Plato in *Charmides* (p.174). It is ethics which helps us distinguish Good from Evil, and thus keep the society in peace and harmony.

Environmental ethics is that code of behaviour which ensures human progress without jeopardizing the ecological balance.

Life Is Unified

Vidya Niwas Misra

I am really very glad that a symposium has been organized in Gandhi Bhavan on such a vital theme like environmental ethics. It is important because it is the last decade of the 20th century, a decade which will lay the foundation of the 21st century. The era of war and conflict is coming to an end. We are entering a century of peace. It is time to solidify and strengthen the roots of peace.

Man has to learn to live in harmony with his fellow beings. But that is not enough. He has also to be in harmony with nature, animates as well as inanimate.

That we should live together in peace should be the basic ethics of all ethics. What is the source of these ethical values? It is the principles like: life is unified, life is endless, and life is interdependent. We must appreciate that if others are dependent on us, we are also dependent on others.

We must realize that if someone needs us, we must be available to him; we must serve others. If I am not able to serve others, then I have no right to get service from others. It is really not the question of rights. It is the question of duties; how much I am prepared to do for others? If others are important to me, then my world expands and becomes larger. And my prosperity would embrace the prosperity of others. That is why Shri Kamal Nathji said that our ethical values make us think whether or not others also get the food that we eat. Even if it is at the thinking level, it does have its practical implications.

The Cartesian thinking has converted society into a factory with just one goal to increase consumption. The result of all this is before us. Great thinkers like Eric Fromm have started loudly thinking that the "Having Culture" would not bring happiness to man. What we need is a "Becoming Culture".

Mrs Indira Gandhi laid great emphasis on environment and she did because environmental degradation was not limited pollution; it spreads widely and others are also affected without having anything to do with it. The developed countries will suffer the consequences of pollution but the developing countries will also not be spared just because they contributed less or nothing to pollution. Hence it is the duty of all developed as well as developing countries to see that the environment is clean.

Man should respect nature; he should take care of nature and do nothing which is not sustainable economically as well as ecologically. Man is a part of nature. He has no existence outside nature. His life is tied to oxygen he breathes, the food he eats, the water he drinks, etc. His all primary needs find satisfaction from nature.

We have to move towards a society in which man-nature relationship is not based on wants but on needs; and on the unity and integrity of all life on the earth. Let me close my address with the hope that we will see each other in each of us and identify ourselves with the life that is one and unified.

More I Have, Less I Am

Ms Nirmala Deshpande

Grave dangers like pollution of air, land and water and shrinking of the resource base of the earth do not augur well for the future of man. The basic cause of all these problems lies in the minds of the people. I feel that something has to be done to purify the minds of the people.

In this context, I would like to refer to Vinobaji's remarks. When he was asked, what is Gandhi? He said science plus spiritual knowledge is equal to Gandhi (S+S=G). We need a radical and revolutionary change in the thought process of the people to bring about changes in the economy, polity and society. It can come only with the judicious integration of science and spirituality.

May I say that spirituality is not always the same as religious practices. It is the essence of all religions. A social structure based on this combination will alone help us to save ourselves and the Mother Earth.

Self-reliance is a good principle but if the danger we are facing is a global one, our action must also be global. Mrs Gandhi pointed out in Stockholm in 1972 that poverty is the greatest polluter in developing countries like India. Unless poverty is removed, we will not be able to have an ecologically sound society. Lord Buddha said, we need *Karuna* (compassion) for poverty eradication. The Government's programmes alone cannot achieve much. All the wealth of the world has to be harnessed and used for this purpose. Sharing is a necessary condition for progress in this regard. This sharing can start from the community, then go to the nation and then onwards to the world. The exploitation of the Third World can be atoned by a compassionate attitude on the part of the industrialized world to share whatever they have with their unfortunate friends. May I also say that the whole concept of a rise in the standard of living is wrong. If we improve our standard of thinking, the standard of living will acquire a different meaning. Happiness does not depend only on the things that we possess. As the great philospher J Krishnamoorthy used to say, "the more I have, the less I am". That is the situation in most affluent countries of the world today. This also applies to the rich few in the developing countries.

Let me recall a case of an American friend (Patricia) who joined Vinobaji's Ashram. She was with him in the long *Pad Yatras*. Once I told Patricia "you ought to have been born in India. You are more Indian than most of us". The reply she gave I can never forget. She said : "It is good that I was born in America. I now look to India. If I were born in India, I would have looked to America."

Toynbee, the great historian, said something which Prof Kothari quotes very often. "At this supremely dangerous moment in human history, the only way of salvation is the Indian way." What is that Indian way? Toynbee has said that the whole concept of conquest of nature is basically wrong. This concept has led to the exploitation of nature and then to the exploitation of man. Now we have come to the thinning of the ozone layer. The Indian concept of reverence for nature has to be given due attention. And there is need for a basic change in our attitude: saving the environment means saving ourselves. Our needs will be satisfied if we have less and less of greed. Greed can never be satisfied even if millions go without food and shelter. Let me also refer to the Jain principle of *Aparigraha*, i.e. we should voluntarily limit our wants. If one is forced to limit wants, one resists. But if one limits them voluntarily, then the happiness pervades everywhere. We have to give thought to this basic principle also.

I would plead for a non-violent change in the process of thinking. If our thoughts change, our life-styles will also change. Let me close with an *Upanishadic* story. A king once heard of a wise man. He asked his minister to find out where he lived. They searched all the towns of the country, but could not find him. When they reported their failure to the king, he said: "Oh fools, you do not know where a wise man lives. He does not live in cities. He lives in the forests in the company of high mountains, and the confluence of rivers." So let us not denude our forests, let us not pollute our rivers, let us not destroy the beauty for our mountains. That will amount to destroying the abode of the wise and the learned men.

Detachment Ennobles Life

M K Machwe

India with a very old civilization and long history has contributed a lot in evolving and developing ethical norms and providing that discriminating wisdom which can decide between *shreyasya* (preferable) and *preyasya* (pleasurable). Between the preferable and the pleasurable, men of wisdom select the preferable. The study of ancient Indian literature reveals that there are two basic facts or assumptions on which the religious superstructure was built. One of them is that the soul is immortal. Body is only a temporary abode. The impact of this belief helped in deciding between *shreyasya* and *preyasya*. If life on earth is only a passing phase in its journey to the final realization of God through *Moksha* then accumulating pleasurable things does not have much meaning. Our scriptures also say that there is something else in the long journey of the soul toward *Nirvana*. This is not the question of only waiting on the platform. There is something more than that. This life is meant for righteousness and *Nishkam Karma* which purifies the mind and soul for the spiritual enlightenment. Yayati learnt it in two lives; while Nachiketa could know it in his tender age of seven. Nachiketa could not be lured by *Yama* to accept what is pleasurable. He wanted to know the secret of life. He could get it because he knew what he wanted and what was good for him. He knew that only a person not attached to transitory things qualifies for such a knowledge. The idea is to remain detached even while having and using things. It is this detachment which ennobles life.

Another concept that moulded the life of the people in India was the recognition that God pervaded everything inanimate and animate. It is He who keeps thing moving. As such no one has the right to exploit anything for one's own benefit unmindful of the interests of others.

Science can hardly vouch for the existence of God. But one must realize that scientific truths and moral truths are different. And they can coexist for a happy and harmonious living. Knowledge and faith in both is needed.

Hobbesian Worldview Is Not The Answer

B K Roy Burman

I fully share the view that we cannot protect the environment unless we look at the life-style of human beings who inherit and use it. In our student days, we used to read about Renaissance (Enlightenment) in Europe and feel proud of it. Now we have discovered to our horror that the enlightenment in Europe was really the beginning of the era of darkness. The concept of the conquest of nature was the product of this renaissance. It looked at nature as the depository of resources, and hence a property to be used and exploited.

The conquest of nature leads to conquest of man. Colonialism was nothing but the conquest of man by man. It was part of European Renaissance or Enlightenment. This led to a worldview in which every one was against every one else. It was the Hobbesian worldview.

Then there are other worldviews which need our attention. They should not be seen in a genophedic manner. There was a Greek worldview which was projected by Plato, for instance, which sought fulfilment in fellow human beings. Gandhi discarded the Hobbesian worldview and sought fulfilment in one's own neighbourhood and surroundings. It is important to look at various worldviews while discussing the question of environmental ethics. The current environmental problems are legacies of the Hobbesian worldview and colonialism, which came on its heels.

Harmony Should Be Our Motto

D N Chaturvedi

As we develop, we have to produce more waste. This waste should not be allowed to go waste nor should it be allowed to be released into rivers which are worshipped in this land of ours. It should be converted into productive like organic fertilizers, cooking gas, etc.

Let us aim at *Dharma, Artha, Kama* and *Moksha* (duty to others, production, consumption and salvation) simultaneously. Live and let live is not enough. Live in harmony with others should be our motto.

We can climb the mountains, get power from the earth, but we cannot create mountains. While taking energy and climbing the mountains, we must ensure that we do not take away the rights of others; nor destroy the gifts of nature, depriving the future generations of their rights of livelihood.

Unfortunately our values are now debased. When we examine the political values, we see only one goal, i.e. to grab power. All through history, queens killed kings; commanders took over the reins of government by force. We cannot live in peace with such values. Then take the religious values: people of different religions fought for supremacy, forgetting that compassion, love, affection, peace, non-violence and human brotherhood formed the foundation of their faiths. We cannot, therefore, depend on religious leaders alone for new values. Today everything, body, mind and soul, is on sale. Intellectuals are on sale. Women are on sale. Money alone is important. So unless our current political, religious and material values undergo change, we do no appear to be nearer the environmental ethics we are discussing here.

In Varanasi, 140 factories release their effluents into the river Ganga. Our scriptures prohibit urination and throwing of dirt in the river. But we are filling the river with these very wastes.

Let us no treat the rural and tribal people as ignorant. They are full of spiritual power and power to love, and the capacity to tolerate all the nonsense that is going on in the society today in the name of rationality, since and technology. We have to learn a lot from them. They do not pollute; they clean. It is we, the educated and rational people, who pollute the environment.

Mrs Zena Daysh

I would like to take up the question of man-nature relationship for a deeper analysis. Although I agree with much of what the previous speakers said, there are issues which need far greater clarity than we have at present.

The view that we need a human framework for nature is not meaningful to me. We need integrated man-nature framework. The language of man is not the answer. The answer to problems of nature are in nature. We have to have a different hearing sound; an entirely different sound vibration. Had we been listening to nature as carefully as we should, we would have not been in the current situation. What we are up to is a much broader base than the technological man's would provides. We need a truly integrated scientific-cum-spiritual base.

K D Gangrade

There has to be a coordinated development of man and nature. Gandhi called it *Sarvodaya*. In my view, we in India need a five pronged policy prescription. First, we should stabilise not only Indian population but also world population by AD 2000. Second, we must halt resource consumption and pollution per head of industrial and agricultural output. Third, we must shift our social preferences from material goods to social services: education, health, nutrition, etc. Fourth, we should divert capital to production of wage goods; minimise the inequities in the country and the world; and fifth, we must limit total as well as percapita energy consumption and income.

S K Agrawal

All cultures have inbuilt customs, manners and mores which help preserve and conserve the environment. How to use various natural resources so that they are not destroyed or even wasted is given in all cultures. We should examine and codify them. And this code should be self-imposed.

It is now clear that the rising population cannot be sustained unless the resources are used judiciously. Nothing is unlimited in the earth including the air and water. Hence we not need to conserve but also to enrich the nature. And a cross-cultural ethical code of conduct will go a long way in achieving this objective.

Choosing the Long- and Short-term Goals

G L Pandit

On a subject of such a global importance as environmental ethics we have to distinguish between the long- and short-term goals. The long-term goal would be to further develop and enrich the environmental ethics. The short-term goals would be taken care of by the ethical values propagated through various educational and other channels.

Our next question is, what are the universal categorical imperatives of these values? Not to tell a lie, to speak the truth, to mediate, etc., are the examples of values which are not tied to any material or social gains. People just value these as a part of their life-style. Correspondingly for environmental ethics, it would be a good example of categorical imperatives that one should do nothing which would violate the law of ecological harmony. Secondly, nature as a whole is valuable as a transcendental presence in the ecology of human consciousness.

This is what I meant when I said about universal environmental categorical imperatives. The conservation of life, for example, on this planet itself is a goal worth following without any consideration of ends and means. The violation of the laws of ecological harmony will mean that no life will be left on this earth. That is why we attach value to life.

The third question is: what mistakes we have committed, whether in India or elsewhere? Pollution, overpopulation, ozone layer depletion, etc., are the mistakes resulting from the style of life and style of development. We have to take the help of science and technology to correct these mistakes. So technology in itself is not bad.

And finally what is the new paradigm we propose to adopt for a better future? We have to constantly monitor everything including science and technology. If each group and society comes forward and suggests new values and approaches, it will help. Given the ends, what are the means available, i.e. hypothetical imperatives? For example, what system of legislation one can devise at local, national and global levels.

> *The attraction of the seer's super sensible world or worlds within worlds is too great to forego, and the states are too high for they may include survival for the planet. Where the modern scientist is baffled by the secret of the life of plants, the seer offers solutions which, however incredible, make more sense than the dusty mounthings of academicians; what is more, they give philosophic meaning to the totality of life.*
>
> —Peter Tompkins and Christopher Bird in the *Secret of Life of Plants.*

3

ENVIRONMENTAL ETHICS AND EMERGING WORLD ORDER

3

ENVIRONMENTAL ETHICS AND EMERGING WORLD ORDER

Opening Address

R P Misra

It would amount to repetition, if we relate here the story of the dominant worldview of the 20th century and the current division of the world into rich and poor. A recapitulation will, however, not be out of order.

The present world order is the product of the Industrial Revolution, which in turn is the product of European Renaissance and Enlightenment. The changes which have taken place in the material standards of life are unprecedented. True, there were peoples and cultures which reached great heights earlier. The field of philosophy, the old continues unmatched. In the fields like dance, music, painting and crafts, there is not much that we have added recently. And wherever something great has emerged in the sphere we call culture, it was not because of the scientific knowledge, but in spite of it.

What distinguishes the last two centuries from the rest is the great breakthrough in science and technology. Man has been able to subdue nature, at least temporarily, and in the West, to benefit from it materially. All the resources of the earth, energy resources, mineral resources, land resources, and even human resources, were there earlier too. But they could not be used on a scale and to an extent as in the last two centuries.

The change really occurred when Europe leap-frogged the Platonic worldview to adopt the Cartesian worldview; when it discarded synthesis and unity of knowledge to adopt analysis and division of knowledge into unitary disciplines; and when it put reason outside the bounds of Ethics and Morality.

The Renaissance and Enlightenment which Europe witnessed in the 18th and 19th centuries was in fact a grand celebration of the advent of new power that came in the hands of man. Possibilities appeared to be unlimited. The prospects of a global society with West Europe as its centre and the rest of the world as periphery became real. Thanks to the newly acquired scientific knowledge and its applicatory version, the technical knowhow, Europe did unify the world. The tiny countries of this tiny subcontinent acquired colonies some of which had many times more people and natural resources, and were culturally far more advanced. Europe had the guns that others lacked.

Since then, a lot of changes have taken place. The colonies have been liberated; they fought and won their freedom. The gun culture reached even those who were subjugated. And the Western worldview now prevails all over the world: developed or undeveloped, rich or poor. East or West, North or South.

One good thing about knowledge is that it cannot be hidden. Same applies to technology. One can postpone but one cannot stop its spread and diffusion. So the guns and knowhow spread fast, thanks to the two world wars. And thus the colonies disappeared as fast as they had appeared.

It was hoped that a better world order would emerge from the ashes of the colonial empires. But it did not. The new leaders of the erstwhile colonies saw in liberation an opportunity to illuminate their countries with the European brand of Enlightenment even though late by centuries. Today, all the countries of the world are racing towards material prosperity, some through industrial expansion while the others through the exploitation of natural and human resources.

The world order that we have today is still colonial but in a different sense. The colonial empires are gone, but the economic power that the imperial countries acquired by exploiting the world resources to their advantage continues with only marginal difference.

Look at the per capita income, per capita energy consumption, per capita use of mineral resources, per capita consumption of food, or any other aspect of life, our world continues to be divided between the rich and the poor; between the powerful and the weak; between the developed and undeveloped. The division is not only a global phenomenon, it is also national, sub-national and local phenomenon.

Those who have want to have more or at least keep the present edge they have over others. And so the exploitation of nature goes on unabated. And those who do not have want to catch up with those who have. Unable to prey on natural resources elsewhere, they are devouring their own resources. The cumulative effect of all this is overexploitation of nature and human resources all over the world. The overexploitation of nature has produced environmental problems and that of the people has produced social tensions and economic disparities.

Environmental problems have now acquired serious proportions. It so happens that no matter what the source origin, they affect everyone everywhere. So long as environmental pollution it within the tolerable limits of nature, it had to perceptible negative effect on man. Nature has its own mechanism to clean pollutants and distortions in it. Unfortunately, however, the pollution level crossed the limits long back. Now it has reached a stage where the survival of life itself on earth is no longer guaranteed.

The ozone layer depletion, the green house effect, the climatic change, the rise of the sea level, the loss of biodiversity, etc., are real and immediate dangers. When they come, they come with a bang. And then we will not be able to retrace our steps because it would be too late. However there is still time to act, time to redirect our energies and to evolve a style of life which will be sensitive to ecology and environment on the one hand and human welfare on the other. The present world order does not promote either of the above two values. It is highly polluting and destructive of nature and it is highly inequitous in terms of distributive justice. What is worse, the beneficiaries of this order do not want of change; they want to maintain their status. They feel convinced that science and technology which produced these problems would also solve them.

Recent attempts at the global level to arrive at honourable compromises have given only limited results. However, there is recognition of the seriousness of the problem; there is participation in the dialogue to resolve the problems; and there is urge to solve them and share the burden. The urge has, however, not yet taken the form of commensurate global policy decisions.

The June, 1992, United Nations conference on Environment and Development at Rio de Janeiro and the accompanying Earth Summit and NGO meets and projects are signposts of hope. The present world order cannot continue. The old one cannot be dismantled in a day. War cannot be rebutted by war. Its rebuttal lies in peace. The Cartesian worldview can no longer be sustained. It has to be replaced by what one may call a Gandhian worldview. But if Rome was not built in a day, peace too cannot be brought in a day.

The ethical foundations of a peaceful world just do not exist; nor their exist values which can protect the environment. Science and technology have brought about a global society in physical terms but man is still in the grip of divisive forces. His tunnel vision has not changed; and he is yet to come to terms with nature. Perhaps the seeds of a new world order are already sown; the sprouting has begun but the soil lacks the nutrients essential for growth. The soil is a new philosophy of life and the essential nutrient is the Environmental Ethics.

Towards Global Environmental Democracy

Anil Agrawal

At the international level there is an open conflict between the economic and ecological systems. The economic system is greed based. Whatever knowledge exists is being patented. It belongs to someone. It is Intellectual Property. No one use it. But when it comes to ecology, we talk of sharing. The way we are using the ecological system, whether it is land, air, water or plants, is irresponsible and exploitative. A few things about environment have now become clear. I attended the 1972 conference at Stockholm as a journalist. At that time, I did not understand much of what was being discussed there.

The *Chipko Andolan* gave me an opportunity to write on the environmental problems of the mountain areas and the reaction of the people to the growing crisis. The ladies who were spearheading the movement said they did not have legal rights to forests but they do have ethical rights. They had been using the forests for centuries. They could not exist without them. They considered it their moral duty to protect the forest. But behind the moral rights and duties, there was the economic concept; there was the issue of survival. None of the ladies said they would not touch the forest. They wanted to use the forest without destroying it. They had a vested interest in sustainability of the forest. Outsiders had no interest in sustainability, they were interested in timber. The local people may make mistakes and do harm to the forests, but they will learn the lesson soon and take corrective steps. Governments and contractors have no respect for sustainability. Japan cuts forests elsewhere, but 70 per cent of its own land is kept under forests. The work in Sukmajari village in the past 10 years as a social response to the ecological crisis has come up boldly. Priyaratna Misra who launched the work in Sukmajari is now working in Bhosaria and Palamou in Bihar on similar lines. Anna Hazare's work in Ranigaon Siddhi in Ahmednagar also deserves attention. We do not yet know how the *Adivasi* women are coping with the feeling of trees in Baster, Nagaland, Uttarakhand, etc. We don't have a proper information system to know about such events and developments. These may be small activities but together they are important. In Chhotanagpur, the *Adivasis* are trying to protect forests in 80 villages or so under the auspices of the Panchayat system. They don't allow anyone to cut forests.

Some resources can be managed only at the community level. It cannot be done at District, State or National level, because it needs direct participation of the people. Apart from agricultural land, all lands belong to the government. Panchayats have come up in some places. Community environmental democracy comes to the forefront in all such situations.

Let me give an example of the urban sector. In Karachi, Akhtar Hamid Khan, along with 5,00,000 people, involved in developing a modern sewerage system. They have formed Gali Republics. People decide the alignment, cost etc. Can something like this not be tried at the global level?

Let us take the ozone layer depletion. Technology intervention affects all countries. Ninety per cent of the cars in the USA have air-conditioners which release CFC. CFC affects the ozone layer; it also creates global warming through Green House Effect. In a real crisis situation one-fifth of Bangladesh will be submerged. If Bangladesh wants to set up a thermal power plant, the World Bank would not give loan. But there is no pressure of the World Bank on the USA to remove air-conditioners so that the CFC emissions are reduced.

All developing countries are being restructured to open up their economies. All the G-15 countries meeting in Caracus excepting one was restructured. The IMF and the World Bank are US-controlled. If we talk of global solidarity, we should think of a system which has equitable checks and balances and which allows the growth of a country without harming the interests of other countries. Bangladesh should be able to save itself. In the same of environmental solidarity, some countries should not remain in poverty and helplessness.

I quarrelled with Maneka Gandhi because she wanted to sign the climatic change treaty without adequate safeguards. I told her that she would have to sign 10 more such treaties. The international laws that are being framed in the name of global warming, biodiversity, etc., have the same infirmities. China and perhaps India together do not release more than two per cent of the CFC. Ninety-eight per cent of it comes from the developed countries. And yet the developing countries are being asked to share the burden equally. The USA is not ready for a global warming treaty because if its companies have to bear the new burden, they will become uncompetitive. Therefore, all should jointly save the ozone layer! And if any country does not agree to share, then it subjected to trade sanctions. What sanctions can we resort to and against whom? So today those who are the main culprits of environmental deterioration want to control those who are lesser culprits! But that is the reality.

The world system in which all have equal rights is not being talked about in UNCED. We must aim at global environmental democracy. We must become citizens of the global community. When we consider this whole earth as ours and our responsibility, then alone we can develop a truly global citizenship.

These are some of the ideas which are being brought forward by NGOs at UNCED. Whatever I have said is part of the art of the desirable. It is not part of the art of the feasible. When we talk of the art of the feasible, we have to talk with different parameters and ask the question: how much ecological sovereignty we are prepared to forgo?

UNEP and World Meteorological Organization some three years ago constituted an Inter-governmental Panel on Climate Change with scientists from all over the world. They concluded that the percentage of certain gases like carbon dioxide, CFC, methane and carbon monoxide in the atmosphere has increased and it could lead to global warming. They have, however, pointed out several uncertainties. Is the increase because of natural or human causes? If because of natural causes what can we do? Will there be global warming if the increase continues? What will really happen in the event of global warming? Would it be good or bad? Would the good and bad effects be equally distributed? If the Tundra ice melts, the exposed moth will release methane gas. What will be its effect on clouds? It may help the dry zones. But they also concluded that if global warming does take place, it will be too late to reverse it. Therefore, as a precautionary step, we should reduce the use of all those resources which release gases that can generate the Green House Effect. The basic point is that irrespective of uncertainty, there is likelihood. Let us not live under the threat of our disappearing.

In any case, the efforts to control emissions are desirable for other reasons also. Protecting forests, energy conservation, reducing carbon dioxide or CFC emission are desirable even otherwise.

Environmental management is a preventive measure for this purpose. But the polluting factories want time to evolve effective measures. Environmentalists however, say when it is known that something is toxic, why should they need time to take necessary steps? Green Party and similar organizations are spearheading the case of environmentalists.

Development Without Destruction

B M Shukla

Shri Kamal Nath gave figures showing that if the developed countries reduced pollutants by 20 per cent, the per head emission of CFC in the world would be reduced to four tons per head. At the same time if the developing countries increase their emission by 20 per cent the per capita emission will come to just 0.6 tons. The gap is enormous. The contribution of CFC by the developing countries when seen in per capita terms is very small. So the problem of ozone layer depletion is not the creation of the developing countries. It has been created by the industrialized countries. It is, therefore, the ethical responsibility of the developed countries not only to reduce the use of pollutants but also to help the developing countries with resources and technologies which reduce pollution. But, instead, they are trying through international conventions and treaties to block access to technology under the pretext of Intellectual Property Rights (IPR).

Let us now take the case of NPT. So much pressure has been put on India to sign this treaty. I call this pressure unethical, because it is being wielded by countries which have acquired nuclear armaments which can destroy the world several times. And they are still resorting to nuclear testing. It is something like the devil quoting scriptures.

It takes my mind back to 1969 when I attended a 3-day seminar on environment organized by the US Aid Mission at Kathmandu. For all the three days, our Nepali friends were lectured in great eloquence on the beauty of natural life and why they should not ape the West and not industrialize. It would be a loss to the world if Nepal became an industrialized country. Some of us then intervened to say that we ought to strike a proper balance. The question need not be posed in either-or terms. It should be in terms of a proper mix.

Should there be development? Development presupposes use of natural resources. It does produce changes in environment. But we cannot allow people to remain poor because the rich people are adversely affected by environmental pollution. What is needed is a shared concern. Those who have too much must scale down their consumption and those who have nothing should be enabled to raise their living standards.

The issues of global warming and ozone layer depletion are important. They will affect adverselly all life on earth. The very possibilities of these things happening should be sufficient ground for action. But such concerns become meaningless in the midst of mass poverty in the develop-ing countries, and lay the ground for doubting even their scientific bases.

We should have a balanced outlook. We should avoid mistakes committed by the West. We should stop vulgar waste of natural resources. We do not know how much courage the leaders of the developing countries would muster at Rio to carry conviction with their counterparts from the West? Any global convention or treaty that we sign at Rio must be equitable; it must ensure sustainable development of the world with emphasis on economic growth in the developing countries and on social development in the developed countries.

Bridging the North-South Divide

C P Singh

In the world of today, most dominant countries are the USA, West European nations and Japan. They are dictating terms politically and otherwise. Rio will also see the domination of that group. The Developing countries, on the other hand, have another point of view. They have been exploited in the past. They do not want to be exploited now. Humanity is divided into two broad groups.

We have examined the implications of development models and arms race. We call for cultural enrichment. We are, however, forgetting that there are other forms of life on earth. They too have rights.

We should speak for all living things in Rio, not only for man. Even wilderness should be protected; Misraji rightly put the question: "Can a river go to the court." No satisfactory answer has yet come from the Symposium.

While looking for light from ancient Indian scriptures, we have to see the whole Indian tradition of 5,000 years; its march from perfect harmony with nature to the present disharmony. There are still a lot of people who will not eat without first feeding animals and other needy human beings. How to share with all other beings without destroying the ecology is the real question. But there is carnivorous life also. Equalitarianism, therefore, cannot be achieved. It is the human beings, as supreme creatures, who should mediate with the system to bring about a balance. Environmental ethics provides the standards for this mediation.

Mrs Zena Daysh

Each one of us present here has to initiate change. And then each institution has to play its role. The reality of politics is that dignitaries like Prime Ministers and Heads of State take decisions in very complex situations. How to make a point there? It took us 16 years to get ourselves into the final communique of the meetings of the Commonwealth Heads of State. We went through four educational seminars. The game is political. We have to make headway with politicians.

4

LAW AND ENVIRONMENTAL ETHICS

4

LAW AND ENVIRONMENTAL ETHICS

Opening Address

R P Misra

Ethical values and jurisdiction provide the framework as well as basis for human action. When the two coincide, the results are optimum-culturally, economically and ecologically. Although morality is one of the fundamental sources from which the two are derived, they approach the issues from two different, even though not always antagonistic, perspectives. The law is enacted and enforced to protect the rights of individuals and aggregates of individuals, while the ethical values are there to ensure that each one performs one's duties towards other individuals and institutions maximally. The difference in approach, at times, brings ethical values and laws in conflict with each other, probably because the source of ethics is the heart, whereas that of formal law is the head. This disjunction is the source of much of the troubles of the modern world. This conflict is more pronounced in matters of ecology and environment.

Overemphasis on the legal rights of an individual viz-a-vis society at large, on the one hand, and the nature's delicate balance for the survival of life on the planet earth, on the other, has brought humanity to the brink of an ecological disaster. Those who are powerful use their rights derived from the concept of "freedom of the individual" to exploit the weak. Racism, slavery, colonialism, etc., are the products of the culture of rights. The earth has been systematically raped all over the world, as nature became a commodity to be exploited and consumed. One has produced poverty, violence, ignorance and premature death; the other has produced ozone layer depletion, Green House Effect, desertification, deforestation and a series of other ecological problems. Modern economies devoid of ethics are not only unsustainable, but also disastrous.

Humans are part of the nature. They can manipulate nature and ignore its laws of balance only to the extent that the life-sustaining capability of nature is not destroyed. Modern man, armed with modern science and technology, has tried to exploit nature to an extent that its life-sustaining capability is jeopardized, endangering the very survival of life on earth. Does an individual or a nation have unfettered rights to exploit nature? If an individual does not have the right to commit suicide, does he have the right to deprive others of life? Each individual on this earth has an equal right to live on earth, but he also has the duty to see that he does not destroy nature, nor does he use more than what legitimately belongs to his share.

To ensure the rights of all individuals, rights must be balanced by individual's duties. In fact the rights and duties should be seen as two sides of the same coin. What is the mechanism one can think of for such a balance? May be, the concept of the rights of the nature would be of help. Nature has the right to self-preservation. Can it go to the court of law? Can the river Ganga, for example, sue the industries and cities which pollute it? This is a moot legal question. The idea has been partly operationalized as public interest law, but it still has a long way to go to be really effective.

Modern age is the age of formal laws. All relations tend to be contractual. Can there be a contractual relation between humans and nature? But even if one can do so, there are problems. All human relations cannot be fully contractual. Most of them are based on ethical principles. The relations between nature and human beings also can not be fully contractual. Most of them will continue to be ethical. When human societies were space-bound and local, not global, man-nature relations were governed by sets of ethical values. Most of those values are now gone. But there is need to resurrect them, may be in new forms and in conformity with the current state of cultural progress. Unless the laws relating to nature are backed by a set of values which do not see man as enemy of nature, but as a part of nature, the current environmental problems will continue to grow and intensify in future.

The point that emerges from the above is that humans have the inalienable right to use nature to save themselves from the jaws of death. Beyond this the right is tempered. It can be enforced only within two ethical obligations. Firstly, people do not use more than their share of the resources and, secondly, they give back to nature that which maintains its life-supporting delicate balance. Gandhiji said that so long as man used nature to meet this needs, no harm was done. But the moment he used it to satisfy his greed, he became destructive. Unfortunately, much of the modern law is based on premises and concepts of an acquisitive, permissive and consumerist society. All these promote greed. There is, therefore, an urgent need to evolve life-styles and development models which are not Greed Promoting, they are Need Satisfying. In this model, development acquires a new meaning. It essentially means cultural enrichment through a judicious balance between economy and ecology and between law and ethics.

While considering the question of environmental ethics in relation to environmental laws, one should not see them as opposite points. They are divergent points needing synchronization and integration. In this context, one may say with some confidence that the process of integration has already begun. More and more of international as well as national laws are tending to conform to the ethical values that protect the nature without jeopardizing the right of the people to use nature for survival and cultural enrichment.

Clean Environment is a Human Right

Satish Shastri

Do inanimate natural objects have the right to go to a court of law? This question was raised by Prof Misra. I am reminded of some decisions of Indian and American courts in this regard. So far as the American court is concerned, it was the Siera Club Purchase case where Mr Justice Douglos delivered a minority judgment that inanimate objects like hills, water-falls, roots, vegetation are as good as human beings in the eyes of the law. They have as much right of protection as human beings.

Similarly, there was a case of the Tennessee Valley Authority where a dam was being constructed by the federal authority in 1967. Ninety per cent of the project was completed. A scientist found that there was a fish named Snail Darter which was found only at the dam site. If the water was impounded, the fish life may be extinct. A case was filed by a scientist against the authority. It went to the Supreme Court and it ruled that the project be stopped. The President overuled the judgement on the ground that adequate measures would be taken to protect the fish.

In India, in the Silent Valley case, the Kerala High Court did not help the environmentalists even though it was clear that many virgin forests with rare species of vegetation and rare animals would be wiped out. Mrs Indira Gandhi had to intervene and the project was stopped by the government. It was approved only after the project was scaled down substantially. Everyone including inanimate objects have the right to go to court. Law does not provide these rights, but judicial authorities like Supreme Courts do take cognizence of the rights of all including inanimate objects.

With regard to environmental ethics, I think we have to have a fresh look at our cultural values and moorings contained in our scriptures. Living in harmony with nature is the recurrent theme of Indian culture.

That life is one and united; and that a single thread connects all life has to be the basic principle on which human society is built. Unity, equity, peaceful coexistence and non-violence should pervade every walk of life. Human body is formed by *Panch Bhootas* which are inanimate. How can we then delink ourselves from inanimate nature. We are not outside the natural forces. We must not destroy the nature to satisfy our greed. We must behave non-violently. If we behave violently nature too will behave violently. This is our ancient environmental ethics.

I will now refer to just two provisions of the Indian Constitution which reflect ancient Indian tradition. Fundamental Rights in the USA include the right to clean environment. Sixteen American States have adopted this law. In India we have put it as Fundamental Duty, i.e. duty to protect and improve the environment. Duty to protect environment takes us to equity also. Similarly Article 48A inserted under the 1976 Amendment enjoins on all States to protect the environment. Law is the manifestation of the will of the people. A Welfare State will always protect the environment.

Rights of Natural Entities and Future Generations

Upendra Baxi

A number of questions have emerged in the discussion. One of them is the question whether entities other than human beings have rights. Does nature have rights? If trees, for example, have rights, then against whom these rights are to be enforced? Law and ethics presuppose what the duties of the present generation towards future generations are. Both these issues, rights of other natural entities and the rights of future generations, are important issues to think about in the university set-up.

The question of intergenerational responsibilities is a difficult one. Why should we regard human survival central? If we can argue that nature will survive without human beings, we have strong arguments for genocide.

This two questions have been bedevilling me. We must resolve them satisfactorily. Then alone we can evolve a meaningful code of environmental ethics.

K R Dikshit

On the two conceptual and philosophical questions (1) whether man is responsible to future generations and, (2) whether nature has legal rights,

I think we need not worry about other responsibilities so long as we remain committed to the welfare of the present society. We would increase our responsibilities several fold if we make ourselves responsible to future generations.

About the second question, I would add that there is undoubtedly a biological unity that binds all life. But we know that plants alone produce; all other living species are only consumers. Even if man does not interfere, there are violent interruptions in nature. In the past, climates have changed, sea levels have risen or gone down and earthquakes have occurred. Nature returns to some equilibrium even after great upheavels. Some species do die. New plants grow adapting to new situations.

If nature has the same rights as human beings, then animals and human beings don't have the right to use plant, the primary producer. If we accept this thesis then there will be no life on earth. Only those which use solar energy directly have the right to subsist.

I do not agree with the view that all life on earth has the same value and hence is equal. We should be rational. We should not overexploit nature. But we have to recognize that human beings have the right to live on other forms of life on earth.

Environmental Laws in India

P S Sangal

There was a provision in the Indian Forest Act, 1927, which permitted the State Governments to transfer land from forest to non-forest uses. In 1980, forest land was put in the central list of the Constitution. There were similar legislations about water, air, wildlife, etc.

These legislations, however, were not considered adequate. An umbrella Act was needed to cover all resources. The 1986 Environmental Act was, therefore, enacted. Unfortunately, it has not achieved its purpose as it was enacted in haste. The Delhi University law teachers met Mr Sheshan, then Secretary of Environment, to tell him how one provision of the Act cancelled out the other provisions. The public Liabilities Insurance Act, 1991 was enacted to help Bhopal victims. Under the provisions of this Act, the company without any examination will pay up to Rs 25000. It thus makes provisions for no fault liability, i.e. simply because you own a factory you have to give immediate relief irrespective of the cause of accident. Owners will be entitled to take insurance under the provisions of the Act.

Another key legislation was enacted in March, 1991. It relates to the labelling of environment friendly products. Some companies have already started eco-marking.

Courts, particularly the Supreme Court, have been very helpful in matters environmental. I may even say that the Supreme Court of India has become a very potent force of environmental protection.

The concern for environment in India has increased since 1972 when the world community met at Stockholm to discuss the environmental problems afflicting the world. There were provisions to protect environment even before 1972. But since then many laws have been enacted as a follow-up of the Stockholm recommendations. Some of these laws are:

1974 — Prevention of Pollution of Water Act.
1976 — 2nd Amendment to the Constitution providing for environmental protection as a Fundamental Duty of a citizen.
1978 — Prevention and Control of Water Pollution Act.
1981 — Prevention of Air Pollution Act.

Laws Alone Will Not Help

B M Shukla

Now coming to the legal aspects mentioned by Prof Sangal, I am convinced on the basis of past experience that passing of more laws is not enough. A certain amount of success will certainly come but if the Bhopal tragedy is an indicater, such laws and inspections may not provide foolproof safety. Just 15 days before the tragedy, the Government of Madhya Pradesh had cleared the plant as safe. The records of pollution control offices show that they can be easily called pollution-causing offices. In the Municipal Corporation of Bombay, there was a case of location of a polluting chemical factory. Scientists and people opposed the project and won the case in court. But the government appealed against the judgment and got permission from a higher court to start the factory.

Prof Misra and Prof Baxi raised some very important questions. I feel that the rights of natural entities are there only in relation to human rights. Human beings are a part of nature. They can use nature only to the extent that it continues to sustain humanity. And hence there are limits.

As regards intergenerational responsibilities, I think these responsibilities are instinctive, philosophical and biological at the same time. They cannot be confined to legal theories alone. I am sure the Symposium will discuss these issues in greater depth. Finally, I would assert that laws and rules are followed only if they are rooted in the cultures and ethical values of the people. India has a special philosophical outlook favouring environmental protection. If this country is going to make a mark in this respect, a deeper look at its cultural heritage is essential. Unlike the Semetic religions, oriental religions have great regard for nature and life. Starting from the Vedas, all the way to the present, respect for the animal kingdom, mountains, rivers, etc., is capsuled in the concept of *Vasudhaiv Kutumbakam.* These aspects were emphasized by other speakers also.

The ancient Hindu philosophy is rich in environmental ethics. We must look back and find out the relevance of ancient thought to modern problems. Gandhi was a continuation of that great tradition. Let us recapture it so that our children do not blame us as reckless and irresponsible ancestors.

R P Misra

We do need formal laws to manage complex social systems. But these systems are not governed by laws alone. There are ethical values which guide individuals and groups. These values and the formal laws must work in tandem. The closer they are to each other, the better it is for the society. The world will be happier if formal laws and ethics are supportive of each other, rather than working at cross purposes.

5

TRIBAL AND RURAL EXPERIENTIAL KNOWLEDGE AND ENVIRONMENTAL ETHICS

5

TRIBAL AND RURAL EXPERIENTIAL KNOWLEDGE AND ENVIRONMENTAL ETHICS

Opening Address

R P Misra

Annihilation, deprivation and marginalization of the indigenous, tribal and rural population is one of the by products of the colonial world order. It was neither intentional nor designed. it was inherent to the very process of development it initiated and promoted.

Rural settlements all over the world were the real producers of goods and services on a scale which rarely produced ecological changes beyond the corrective limits of nature. Some of the finest cloths and handicrafts were produced in tiny villages and small townships almost everywhere. The spices and silk for which India was so famous were not produced in cities. Villages were the real centres of production. They were tuned to local ecology, local resources and local culture. What they produced was unique, and place and culture specific.

There are cities which functioned as commercial, administrative and cultural centres. They were not industrial centres and there were limits to their size as well as growth. Population growth was also slow due perhaps to lower birth rates and higher death rates and hence the pressure of people on land was lighter. There were vast areas of wilderness which provided shelter to a great variety of plants, animals, and human beings. These areas were witness to a cultural panorama of great variety people living in different stages of cultural evolution and presenting varied models of life and living. Only the most accessible lands were inhabited and cultivated.

Even the regions marked by complete villagization had wildlife preserves. There were village forests, village pastures and village commons. The economy of the village was integrated with its ecology. The *vice-versa* was also true. There are changes indeed. But such changes were slow. They were rarely rapid and violent and hence were absorbed and internalized by the system.

This slow moving but highly integrated life and living was not all that rosy. Man's creative faculties were very much circumscribed by the overriding power of nature. The death rate was indeed high. Lots of people did not have a decent living. The rituals and conventions which took the place of philosophy, reason and knowledge held human progress in bondage. Yet, in spite of all these or perhaps because of these, man never thought of conquering nature.

The world of an average individual in such a society was limited to a radius of a score miles though long-distance travel was not rate. It was a pedestrian, horse-buggy, bullock-cart, horseback, and wooden-boat society. Slow movement made community bonds stronger. People knew individuals, families and villages in great detail. They also knew the local nature and culture in depth. People were in love with their environment which sustained them and shaped their life. Each tree had its significance; each animal had its importance. Nothing was useless. It was a society of use values wherein everything had its place.

There were scientific discoveries and technological breakthroughs, perhaps not as spectacular as in the Cartesian era, but highly significant for that time. But many of them were never made public. Writing about submarines, Leonard da Vinci wrote in his *Note Books*, "How by an appliance many are able to remain for some time under water. How and why, I do not describe the method. This I do not publish or divulge on account of the evil nature of men who would practise assassination at the bottom of the sea in their lowest parts and sinking them together with the crews."

John Ruskin, in *Modern Painters*, said, "No changing of place at a hundred miles an hour, nor making of stuffs a thousand yards a minute will make us one whit stronger, happier or wiser. There was always more in the world than the men could see: they will see it no better for going fast. As far being able to talk from place to place, that is, indeed, well and convenient but suppose you have originally nothing to say! We shall at last be obliged to confess, what we should have long known, that the really precious things are thought and sight, not pace. It does a bullet no good to go fast and a man, if he be truly a man, no harm to go slow; for his glory is not at all in going but in being." The pre-Cartesian man was "Being". He was in harmony with the environment, both natural and human.

The present-day rural and tribal people are the remnants of that pre-Cartesian society. But they do not represent them now. The Industrial Revolution has distorted them. They have been under pressure; their wherewithal in undermined, and their ecological setting is modified, disturbed and destroyed. Yet the remnants of once an integrated community which could sustain for centuries a fairly high level of culture are still there. Like the virgin and pristine forest stands, there are still some pristine communities. They are no longer a model for our future but they provide living laboratories to understand how man can develop without destruction. Such communities should not be marginalized to the rugged hills and denuded lands; conditions should be created for their self-reliant development integrating the past and the present into a new future.

Of particular interest to us are the ethical values of these communities; values pertaining to nature, individuals, groups and other forms of life; values about development; and about self. They have lived nature and many of them still live, in contrast to the modern urban society which outlived nature long ago.

The new social order in the offing calls for global thinking and local action. One is not sure of the shape this order will take. From what one can see from the bits and pieces is that it would be far more humane and scientific at the same time. The local action has by necessity to be based on local knowledge and local values modified and refined by global perspectives. The rural and tribal societies not only provide the starting point for action but also for learning from the people who have preserved something which others have lost.

Rural and tribal knowledge is not created in distant laboratories and then brought to the user. It is time-honoured knowledge of millenia which oftentimes approaches the depth and pervasiveness of wisdom. It has come from the experience of the people, from their laboratories of time. This experiential knowledge may be out of date when seen against the scientific knowledge of today, but it is benign without side effects and hence futuristic too.

People's Perception of Nature

B K Roy Burman

In 1954 an Amerindian chief wrote a letter to the President of the USA to say something like this: "You the Chief of the White want to buy our land. You want to purchase our land. But please remember, that the trees, the rivers, and the birds are our brothers and sisters. Look after them. We bequeath that trust in you." But what has happened to that trust?

I happen to have some knowledge of tribal society. Some four years back, 10 villages spread over the hill areas of Nagaland, Manipur and Assam were studied by a team of anthropologists, economists, sociologists, chemists, geographers, agronomists, physicists and engineers. Our idea was to study their perception and attitude towards nature and then see how it deviated from what we considered to be a rational approach.

We collected about 1,500 plant specimens. The tribals could identify more than 700 specimens. We included one non-tribal village in Manipur valley. This village also had same knowledge about plants. The university botonists, agronomists and foresters could identify only about 400 specimens, while the tribals could identify more than 700 specimens. The tribals claimed to know five varieties of plants which were insect-repellant. This claim was found to be valid (after laboratory examination). We asked our friends from development departments if they knew about those plants. The answer was negative. Millions of rupees have been spent in buying insecticides. But the remedy available locally was not known to the extension officers. We are importing ignorance and marginalizing the real knowledge. The basic document of the Rio conference also states that indigenous people are the custodians of knowledge of mankind. Quite much of the knowledge of medicine is derived from the indigenous people like Amerindians in Americas. It has been put in the market after slight processing. People who speak about the intellectual property right today are the ones who have stolen this knowledge. What right they have to talk about IPR? Whom does this right belong to? Here is a substantive ethical question.

There are some who think that the endowments of nature cannot be allowed to go waste; they must be used for human welfare. But is the sublime height of the Himalayas a resource for utilization? Can we value it in terms of money? Great poems have been composed at the inspiration of the Himalayas. Does a literary work have a price tag? The desolate Himalayan heights are barren and waste but without them Indian would have not been what it is both physically and culturally. We can perceive the value of the Himalayas only when we throw away the non-ethical and non-human view of the world lock, stock and barrel.

The other day I heard experts talking about wasteland development. I asked them: is there a wasteland? What we consider waste is the resource of the poor. The poor are being evicted from this land which once belonged to them. And our elite call it wasteland! Wasteland is the product of the minds of those who can impose their will. It is the perception of those who wear the blinkers of modernity and are highly imbued with the feeling of self-righteousness.

Anything that is not assigned by the sovereign, the king or the State, belongs to the State. This is a value system that we enherited from our colonial past. Several of our historians like R S Sharma, Irfan Habib, Romila Thapar, and also our honorable members of judiciary like Justice Hidayatullah and Justice Bhagwati have raised this issue. Prof Neelkanth Shastri mentions that there is only one mention in our ancient literature about the king being the owner of the land. Otherwise the king was a protector, not owner. He charged rent for protection.

The State ownership of land was introduced by the colonialists who were charged with the energy released by European Enlightenment. Unless assigned by the sovereign, land does not belong to the people. It is called legal positivism. I chaired a committee of the Government of India on the land holding system of the tribals. The committee had a governor and a judge as members. We found that by the operation of this doctrine of sovereign as owner in one State in the central belt of India, hardly one per cent of the land under tribal occupation for generations was recorded in their favour in the record of rights. It happened on two grounds: (1) application of the concept of *resnoli*, i.e. that which is not assigned to anyone belongs to the State; and (2) only that land which is permanently deforested is assigned to people. Not even the land under shifting cultivation belongs to tribals because such lands are not permanently deforested.

The State has leased out much of the land to private companies for commercial exploitation. The eco-diversity of the forest is being destroyed at a very rapid rate. But who is responsible for it? That is the ethics of the State. Our report was presented to the Parliament but no political party or individual MP took note of it. Tribal people do not speak in Parliament. When they see no hope at all, they become Naxalites. We condemn and kill them because they use violence. But we do not realize that it is the State which used violence against them first. If violence is for justice, and it is resorted to because there is no other means to get justice, then it borders on non-violence. Violence is the only language that governments understand today.

K D Gangrade

Prof Roy Burman has raised the issue of Intellectual Property Right (IPR). His comments apply to all of us who conduct research on people to write papers and books. We do not have the courtesy to even acknowledge the source of knowledge. Knowledge given by the people has no value. It is our interpretation that has value. We have to rethink about our values in this regard. So IPR has to be examined at our levels also.

Towards a Culture of Care, Share, Quest and Bequest

R R Singh

My plea is for a culture of care, share, quest and bequest rather than a culture of conquest and domination. In terms of peasant societies and rural experience, I am reminded of the sayings of Ghagh in Eastern U.P. and Bihar which guide the farmers in their agricultural operations. Many trees are marked with gods and goddesses. They provide shades and shelter for birds, animals and human beings. In this tradition, they are not just objects of consumption. Some of the trees were planted not for getting any benefit in one's lifetime. Tamarind, shisham, etc., were planted for the next generation. These were the values of the rural people.

In this connection, I would like to bring to the attention of the Symposium the values cherished by the *bishnois* of Rajasthan about which I understand, Prof Indra Pal will talk in some detail later on. The Chipko movement is based on the experience of the *bishnois*. Another case of values promoting equity is the *Pani Panchayat* near Pune. Both water and land are equally distributed even though ownership of land varies. Conversion or convolution of intrinsic value to use-value has benefited each and every member of the *panchayat*.

Real crisis before us is the growing conflict between need and greed. Nation societies have rights to fulfil their needs in accordance with their own conception. But do they have the right to interfere with similar rights of other nation societies? The concept of basic needs, therefore, cannot be defined within the confines of national boundaries, if we are thinking of global environmental ethics. The politicization of the issue will go out of the philosophical and moral frame of reference.

Can the developed societies bring down their consumption level to lead a more austere life in this generation? And in the case of developing countries which are consuming less, can improvements in consumption of goods and services continue till poverty is banished?

It seems to me that it would be easier to improve the living standards of the developing countries using the current models of development than to reduce the consumption level of the developed countries in the short run. Several generations will have to consistently aim at such changes to achieve any success. I am reminded of my experience with the Chambal valley dacoits. A deliberate decision was taken by the government after the surrender of dacoits to lodge them in class I cells of prisons. It was not appreciated by the public. But it was a tactical device to provide all facilities so that when they were released after 2-3 years, they would not revert to the rugged and barbaric mode of life again. They were in a way corrupted to lead a useful life.

Many dacoits confessed to me that they were so used to comfortable life that they and no stamina and inclination to walk 25 km in the night through the rugged terrain any longer. So the people who are used to a high standard of material life have a far greater problem in climbing down than we think. If we are going to bargain, we will not succeed. We can neither force them, nor work on *quid pro quo* basis. But if we try with tolerance and empathy to understand the problems of climb-down that the developed societies have to face, then we can make some headway. It may take a whole lifetime to effect any changes; but the change will take place. Let me refer to the contents of a recent UNDP Human Development Report, 1992. A similar exercise should be attempted with regard to environment and development issues. This report talks about (1) human development, (2) human deprivation, and (3) human capital formation. I have pleaded for a similar report on environment with the focus on: (1) environment development, (2) environmental distress, and (3) environmental capital formation.

The traditional code of conduct in village societies gives more to nature than what it take from it. So evolution of a global environmental ethics demands consideration of the country's ethical tradition; existing constitutional and legal supports; current environmental situation; nature of citizens' initiative, anticipated societal demands on environment and its carrying and regenerative capacity to meet them. In fact, in one of the journals *Grassroots Development*, Daly has mentioned about development as an objective. He has defined development as a qualitative improvement of physico-economic base that is maintained in a steady state by a throughput of matter and energy within the regenerative and assimilative capacity of the ecosystem.

M K Machwe

I would like to pose a question. When we are talking about ethics and ethical behaviour, are we talking in absolute or relative terms? Someone could very well say that it would be very unethical if you do not depute me to the Rio conference.

Earlier, Prof Roy Burman commented on wasteland development. He said that it would deprive the poor of their livelihood. But is it not unethical to underutilize the land and thus deprive a larger number of people a better quality of life?

Wasteland Is Not Really Waste

B K Roy Burman

What I said is that land which is a resource for a section of the society is being called waste. This is unethical. Whether it is ethical to put the land to more efficient use or not is another issue. What happens to the livelihood of the people who depend on a given land after it is put to "more efficient" use, is an important question to be considered.

In Tamil Nadu, I saw, last year, *kesuriana* on land which was the source of livelihood for a large number of poor people. These people no longer have access to the land. They have gone to the urban slums. It was a wasteland but a resource for the poor. Absentee landlords who have put *kesuriana* are enjoying the benefits of it. Is it more efficient use? Waste for whom? is the question that remains unanswered.

R P Misra

The disjunction between ecology and economy can be clearly seen from the way we measure GDP. We count, for example, the output of cows (milk, hides, bones, etc.) as well as the value of cows in the GDP. But when it comes to plants, it gets into the GDP only when it is cut. Wood is a resource, but not the plant. Forest gets into the GDP only when it is destroyed. That is how our economy is evaluated. Monetary value of timber gives value to forest. On the one side we are saying that forests are essential for human survival and hence they cannot be evaluated in terms of money; on the other, only the dead forest has any importance for us.

I share the views of Prof Gangrade and Prof Roy Burman about the attempt of the developed countries to pressurize the developing countries to accept a neocolonial world order. This approach and attitude of the developed countries is going to mar the deliberations of Rio conference. Despite disunity among the developing countries, they are not going to be satisfied by lollypops. They have come to know of the exploitative relationship that drains the poor countries of their basic vitality.

The process that is operating at the global level is also operating at the national and local levels. What are we doing to our own poor and marginalized people? What ethical right have we to point our finger towards the developed countries when what we are doing at national and local levels is exactly what they are doing at the global level. If we are prepared to sacrifice anything for our own marginalized and poor people, how can we expect the industrialized countries to sacrifice anything for the less fortunate developing countries?

K R Dikshit

I have some personal experience of working with tribal people. They do know about the plants, they cannot, however, classify them. Most tribals can identify various plants and tell their local names. But I doubt if their knowledge is superior to the knowledge contained in botanical survey reports. Many of us perhaps do not know that the flora of the country is well surveyed. For example, in western India to which I belong, the *Flora of the Bombay Presidency* (by Cook) is published in three volumes. It describes each plant species; its location, properties, etc., are given in detail. Its vernacular name is also given. I have mentioned this in my book *Forest Ecology and Man*. Where I could not identify a plant, I asked the local people. They gave the Marathi names. I consulted the book, and got the Latin name. So, the knowledge of the people about their environment is certainly comprehensive, but not exhaustive nor organized.

Philosophically speaking, the IPR with respect to these plants belongs to them. But it is very difficult to prove and enforce it. A company comes and develops something out of the plant or someone develops an idea and patents it. How to patent the findings of people who are the original discoverers of the plants? It is a practical problem. But I agree that we have to recognize their rights.

As regards the use of wasteland, I do not agree that it is unethical to use it better. The tribals do not use the land optimally because they do not have the benefit of modern technology. For example, tremendous amount of grazing has gone on in Rajasthan for quite a long time. The case of shifting cultivation was already mentioned. There is a view that th soils is poor and hence a five-year cycle is essential. But I believe that with modern technology and knowhow, it can be made highly productive.

I am not saying that the tribal people should be deprived of their habitat. They should have their legitimate share. But to keep the land under low productivity is criminal . I just do not understand the logic of our colleagues in this regard. Let all the land be optimally used. Let me cite the example of a successful project in this regard. In Dang district of Maharashtra 700 ha of land has been scientifically surveyed. The tribals have been settled in villages. They are now using developed land and are happier than before. Through the technological intervention in the form of reorganization or additional inputs, we can improve the living conditions of the people. More efficient use of resources will benefit all.

B K Roy Burman

As regards the knowledge of tribals and peasants about plant life, I reiterate that the botanical surveys are still very deficient. Let me give another instance to prove my point. I was the chairman of a committee of Forests and Tribals with seven foresters as members. Our inquiries revealed that forest surveys had listed only 300 species out of a total of 1,000 species or or so. I asked the Director General of Forest Research Institute, Dehradun, Dr. Tiwari, about the research done on plant communities in India. Like human beings plants also live in communities. Introduction to or removal of a plant from a plant community makes a lot of difference for the whole community. He told me that the FRI had started research in that area only recently. Only about half a dozen studies have been done in the Northeastern States. So this is what we have achieved in the past 100 years in forest research. Foresters' attention was on timber; not on forest.

Many eminent foresters in India like Dr. Tiwari have great knowledge about forests, plants and animals, and people who depend on them. But their knowledge is derived from experience, not from the system of knowledge we promote in our universities and research institutions. Their personal involvement has given them knowledge. The ground is now being prepared for scientific research which can provide the knowledge people possess with regard to shifting cultivation. I still maintain that shifting cultivation is based on sound ecological principles. I say this on the basis of ICAR and agricultural university reports and my own observations. Shifting cultivation induces highest soil erosion when a particular technique is used. But under another set of techniques, soil erosion is the least. The adverse comments of the Planning Commission on shifting cultivation are based on data of land with 60-70 per cent slope. Here the erosion is maximum. But such shifting cultivation is not common in such lands. The positive side of this cultivation is not presented. When confronted with hard facts, they only concede this far that they do not have enough data about shifting cultivation to be conclusive about its adverse effects on soil.

Chipko Movement Is 300 Years Old

Indra Pal

I would like to put before the symposium a few points on environmental ethics of the people of the Indian desert. Rajasthan desert is perhaps the most densely populated desert in the world.

As a student of geography, I know that camel is the most suitable animal for desert climates. But now I also know that *khejri* is the most appropriate tree of the Rajasthan desert. As we know, there is not much of vegetation in deserts. There are many bushes and grasses. In between are a few trees. Among the trees, *khejri*, whose botanical name is *prosopis cineraria* and in Sanskrit it is called *shani*, is the most important.

This tree is identified with a *bishnois* of Rajasthan who are the followers of saint Jamboji. Jamboji stayed, during a severe famine, in the heart of the desert. All his followers had migrated but he managed to stay in the desert by using the *khejri* tree: its leaves, fruits and bark. Every part of the tree has nutritive value. When the drought was over and the people came back, Jamboji asked them not to cut the *khejri* tree.

How the people of Rajasthan internalized his advice to protect scarce plant life is exemplified by what happened in Khejari village some 300 also. In 1720 the Maharaja of Jodhpur decided to get some trees of Khejari village cut. He needed firewood for lime kilns. The people of the village did not allow the felling of trees. Three hundred and sixty people were killed by the soldiers but they could not cut the trees. This was the first Chipko movement in India.

The Central Arid Zone Research Institute (CAZRI) has also found that khejri is a suitable and useful desert tree. It sheds leaves during one season but in general it is evergreen. Desert animals like sheep, goat and camel live on the greens of this tree. Its roots go so deep in the soil that it can survive even in worst droughts. It has been found that some of the trees send their roots to 120 ft in the ground. It does not take anything from the top soil. Under its canopy lots of other plants and trees grow. Unlike the eucalyptus, it improves soil fertility.

The *khejri* tree is integrated with the whole life support system of the desert. This tree is grown in the cultivated land also. Every part of the tree is useful. Flowers have blood purifying medicinal value; it is also used as medicine for miscarriage; it indicates availability of ground water. And women, who do not get a son, hang their *cholis* (brassiers) on the tree and it is believed that they bear male children after this ceremony. No wonder then that khejri is worshiped all over the desert region.

The Rajputs consider this tree as a symbol of power and strength. It is believed that during the *Mahabharata*, the Pandavas hung their bows and ornaments on this tree when they went out incognito to the forest. Even today in the State of Jaipur, the Maharaja elaborately performs *Pooja* on the Dussehra day. Religious scholars have written books, one of 912 pages, on the ritual. The common people also follow the same but less elaborately.

The general practice is to plant this tree on the northeastern side of the settlement. I do not know why? But the question that intrigues me most is: why has this tree not been planted in Rajasthan on a large scale since independence. It is perhaps because our forests and developers think that anything traditional must be discarded and anything modern and imported like eucalyptus must be accepted?

The noted geographer Huntington has written about the Kazaks of Central Asia. They are animal rearers but are vegetarians. If they eat meat, they cannot develop animal husbandry in Kazakistan, a region not very rich in vegetation, says Huntington. Similarly in Rajasthan, the life of the people depends on the khejri tree. They cannot afford to cut this tree. It is a question of survival of the people. Ecological planning not only ensures survival but also ensures sustainable development.

Learning from Others

Ms Aruna Luthra

I think we are eulogizing Gandhi and Indian culture too much. Let us not isloate ourselves from the global economy and environmental concerns. Gandhi was not the discoverer of what is being said about him. We should not have any exaggerated sense of nationalism. I am not trying to run down Gandhi but to ascribe all good to Gandhi serves no purpose.

Open your eyes to the larger world, the world which is equally interested in the protection of the Earth. Why do we need protection. Look at what Germany did after World War II. Look at how Israel emerged as a strong nation out of nothing. Why do we ask for protection in the name of poverty, etc? The world is trying to do something to save the environment. Let us join them and make our own contribution. Let us not say there is nothing beyond Gandhi.

K D Gangrade

I think Ms. Luthra has not caught the essence of the Symposium for she was not here yesterday and much of today. For the last two days we have been talking of what is happening at the global level. We are fully aware of the special circumstances under which Germany and Israel developed. To cite them as examples to be emulated by large countries like India with a colonial past amounts to ignorance about this country. Unlike Ms Luthra, who will like India to merge in global economy, Gandhi advised us to adopt what is good elsewhere but not be swept away by foreign waves. Ms Luthra should try to understand Gandhi before making uncharitable comments. She should try to understand her own country India not through the coloured glasses of Germany and Israel but through a balanced view of life. She has to open her eyes to the fundamental realities of this country.

Hans Poser

Let me correct the impression that the German society is an ideal society deserving emulation by the developing countries. We have many of the problems which you have in India plus something else. We have to solve many big problems.

Prof Burman talked about the need to have a new worldview. The philosophical tradition of ethics gives reasons and rules of guidance. Environmental ethics, therefore, must have theoretical background to introduce a new value of nature. But what sort of nature we are talking about? We certaintly don't want to stabilize the natural systems by putting tigers near villages and towns, because that was their habitat earlier. We are thinking of nature in human perspectives.

We do not have the philosophical traditional solution to the newly emerging problems. Nevertheless we have to start with the long established ethical traditions regarding environment which exist in India as well a sin other countries particularly Europe to find appropriate solutions.

There are problems informing a worldview. Some 10 per cent of the people see nature in a new perspective today. This perspective has to be expanded and deepened. This takes us to questions like:

1. How to prepare the foundation for a new value orientation?
2. How to introduce new values so that society can internalize them and act without fear and on its own?

Culture of Sharing Alone Can Solve Our Problems

P C Patanjali

I would like to mention how the rural people still have so much respect for nature and if they do anything which goes against their environmental ethics, it is the outer socio-economic environment over which they have no control, and to handle which there is no ethics at their command.

The *bhogwad* or consumerist culture has mechanized human beings and produced things which degrade the environment. Man has lost touch with the biological realities; his love for nature is on the wane or is already gone. His attitude has become commercialized. He is not concerned with how his thoughts and actions affect his fellow beings, the society and the nature at large, so long as they produce immediate benefits to him.

I, therefore, congratulate Prof Misra for having organized this conference. There are conferences on economic growth, development, science and technology and their relation to environment. But it is the first time that someone thought of organizing a discussion on the cultural dimensions of the environmental question, i.e. environmental ethics.

Man is the product of five life-giving and supporting elements. It enjoins on him to protect these elements. Instead of living in nature and using nature for his welfare in perpetuity, man started milking nature to the finish. He forgot his past and he forgot his future. He not only started living in the present but also living in the today alone and acting as if there is no tomorrow. This has led to an imbalance in nature as also in human society.

If we look at the rural scene, even today when an Indian farmer gets up from bed, he bows to the Mother Earth. He does it because for him Earth is the mother in a real sense. The urban man, the contractor, the businessman and, let me say, even the academicians assembled here, do not have that attachment to nature. This group, as Prof Misra often says, does not salute nature; it salutes science and technology. May be, the farmer has a blind faith in nature; but do we all not have blind faith in science and technology? To the farmer, nature is the mother; to us, technology is the mother.

I would, therefore, plead for the integration of two blind faiths to produce a rational environmental ethics.

Another thing that I would like today is that the ethical values rural India subscribes to today may not be in the original form. But the foundation of these values is sound. Even a poor man in the village, before he puts the first morsel in his mouth, first keeps one aside for the cow. And at the end he leaves something for the dog. That is the ethics of sharing; the recognition that all life is one, and everyone has a share in what we produce and consume.

Our ancestors said that *Bhajan* and *Bhojan* (prayer and dining) should be performed in a quiet place but now, we eat on the roadside and pray in the humdrum of public *melas*. While eating, we talk of business or gossip about others. Now even prayer has become a business. Go to any temple, Gurudwara or mosque. What is being blurted out on the microphones is neither conducive to *Bhajan*, nor to peace in the neighbourhood.

So a lot of land, water, air, noise and other pollutions are also generated by activities which should essentially be spiritual producing peace and harmony. It is all because of distortions in our values, even religious values, resulting from the invasion of the consumerist culture.

We must, therefore, make ethical education compulsory at all levels of education. We have also to put environment in our curricula from the primary school. I think the Gandhian approach is a must for changing the situation. The voluntary agencies and intellectuals have an important role to play.

6

CULTURE, RELIGION AND ENVIRONMENTAL ETHICS

6

CULTURE, RELIGION AND ENVIRONMENTAL ETHICS

Opening Address

R P Misra

Culture in essence is a way of life. It includes all those forms of behaviour which man learns and acquires from the society and which animals do not possess. Mankind has a double heritage. One comes without efforts. All living things including plants possess it. All human beings find a way to protect themselves, civilized or primitive. All eat and drink. But all that is acquired from learning is part of culture such as language, arts and crafts, science and technology, religion and education.

Religion is a part of culture. In essence it is a body of beliefs, rituals and practices which centre around spiritual beings who possess supernatural powers to protect their followers. All religions, primitive or modern, possess this quality. The so-called higher religions were primitive to begin with. They evolved to the present complex system through centuries of interaction among beliefs, experiences and external influences. The techno-economic advances mark the advent of civilization. Religion and civilization together lay the foundation of a culture.

The world is marked by wide cultural differences. Within the great cultures of today, there are sub-cultures which differ in details but function under a common umbrella of beliefs and values. European, Chinese, American, India and Arab cultures, to name a few, are not as unified as they appear to be. There are powerful sub-cultures within each of them. Each culture and sub-culture is, however, built around a religion or faith and each has its own worldview.

A truly global society must evolve a global culture. It cannot have a global religion, nor it is necessary. If the humanistic elements of various religions are brought together and humanity is able to feel alike on issues pertinent to human survival and development, then the first step in the evolution of a global culture would have been taken. What is common among various great religious traditions with regard to ecology and environment will provide us the clue for the identification and formulation of an appropriate code of conduct for environmental protection, conservation and reconstruction.

Think globally, and act locally has become a universal slogan of the environmental movement. It implies a universal environmental ethic towards which, we feel, we are slowly moving. What are the common principles in that movement. A comparative study of environmental cultures and ethics across the borders of time and space would be of help in this regard.

For this purpose I quote below extracts from *Connect*, a UNESCO publication. Chronologically one might commence with Hinduism, whose origins reach into the dim historical past and resist facile doctrinal definition.

> Hindu thought graviated toward belief in an inner, unseen, abstract reality, underlying the manifest world disclosed by the senses. For man (and other living things) it was *Atman*, spirit or inner self, in contrast to the body. More expansively and abstractly still, all things in the divided, articulate world disclosed to the senses are manifestations of inner Being, *Brahman*. *Atman* and *Brahman*, in turn, are to be conceptually united so that the inner Being or essence of all things is soul, spiritual being, or consciousness. Objective knowledge and subjective knowledge thus coalesce. To know one's self, not one's personality or empirical self, but one's transcendental self, is to know the nature of all things.
>
> The empirical world is important, but unreal, since it seduces the soul into illusion. It distracts the soul from finding itself, attaining liberation, and merging with the one essential, transcendental Self. Other forms of being, particularly other forms of life, are victims of the same deceit, frustration, and suffering as oneself and should be empathied. Indeed, there is no real distinction between self and other. One cannot, thus, profit at the expense of others, either other human beings or other natural, environmental beings, since, ultimately, there are no others all are ephemeral manifestations of one indivisible Being. Lastly, there is a correspondence between the ecological worldview and the worldview of Hindu thought. Ecology also represents the world as a unity, that is, holistically—the unity of oneself and one's surroundings. Thus, there are two major elements in Hinduism which contribute to the development of a universal environmental ethic: empathy and compassion regarding all living things and a sense of harmony with the environment, therefore, its protection and enhancement.

In Jainism, more than in Hinduism or Buddhism, one finds and explicit environmental ethic. Jainism, in contrast to the core philosophy of Hinduism as discussed, is dualistic rather than monistic. There is a fundamental dichotomy between soul and body, mind and matter. Each soul, moreover, maintains its own integrity. It is not a manifestation of the universal soul. Every living thing is endowed with such a soul. And although in each living thing the soul is, as it were, crusted over with flesh and its consciousness dimmed and confused with sensory perception of various modes and degrees of clarity, all souls are equally pure and perfect in and of themselves.

At the moral core of Jainism is the doctrine of *ahimsa*, the determination not to kill or harm any living thing which contains a soul as perfect or complete as one's own and which is as liable to suffering on oneself. The Jains are famous for the extreme lengths to which they go to honour this doctrine. Of course, the eating-of-meat consciousness is more acute than plant consciousness. Mahavira, the founder of Jainism, himself ate only leftover food, prepared for someone else, so as not to have personally caused injury even to the plants, or their seed, from which it was made.

Buddhism stems from the teachings of Siddhartha the Gautama, who lived in India during the 6th century B.C. Buddhism, however, today flourishes less in India than in countries to the south and east of India, namely, Sri Lanka, Myanmar, Tailand, Kampuchea, Laos, Vietnam, China, Korea, Japan, Tibet, and Mongolia.

Core moral values in Buddhism are to be found in the five precepts: abstention from killing living creatures, abstention from stealing, abstention from lying and abstention from taking intoxicants. While these precepts embody the basic requirements for the living of a good life and the establishment of a good community, some of these are relevant to a conservationist ethic. The respect for life and property, the rejection of hedonistic life-style and the notion of truthfulness emphasizing consistency in thought and action are all ethical premises relevant for the development of environmental ethics.

The Buddhist precept concerning abstention from killing living creatures focuses attention on the ethical premise concerning the value of life. Buddha asked people to abstain from destroying the life of human beings and animals and also condemned the infliction of suffering and pain on living creatures. He was also critical of the pleasures of hunting. The kings were expected to provide protected territory not only for human beings but also for the beasts of the forests and birds of the air. The principle of *ahimsa*, non-harming and non-injury to life, was a concept found in the Jain and other Indian sects and the Buddha (though he did not go to the extremes of the Jains) was alive to the concept and preached against talking life. All this shows great feeling of sympathy for living creatures.

What can be inferred from the philosophy of Buddhism is a pro-conservationist (sound management) conception towards nature, which is critical of an aggressive attempt to exploit the environment for short-term benefits and generate gigantism, and a life-style based on limitless consumerism. In short, a non-violent and gentle attitude towards nature, animals and fellow people provides the essence of the environmental stance—the environmental ethic—of Buddhism.

Zen Buddhism provides an especially fitting philosophical and experiential basis for an environmental ethic: the phenomenal world is affirmed as the delightful expression, the artifice and play, of the benign and loving common essence in all things. There is, moreover, a very strong tradition, evident in Zen poetry and art, of a nature esthetic. The contemplation of the fleeting yet eternal moment of satori are all elements of an esthetic attitude towards the environment. And the esthetic value of nature has long served as a powerful human motive for its conservation.

Zen Buddhism has certain affinities with Taoism, since Zen is a version of Buddhism which evolved in China, where Taoism is native. The word tao literally means a way, or a road. It is the way of the universe, the orderly and harmonious unfolding of phenomena, the developmental tendency of things. If allowed to take its course, it results in natural fulfilment and perfection.

Taoism stresses the perfection of harmony between humanity and nature. It also provides the basis for a philosophy of technological development. The traditional Western forms of high and hard technology should be abandoned, from the Taoist point of view, for forms of low and soft technology or what is sometimes called an appropriate technology. An appropriate technology is essentially adaptive and cooperative. It does not attempt to command or control nature, rather, its approach is to bend natural processes to human advantage and adapt human ways of life to the environment.

Like Zen Buddhism, Taoism conceives the environment as an articulate unity, a unity among natural things and these things with humanity. This picture of nature as an autonomous and dynamic whole, in which humanity has its fitting and appropriate place, fit well the worldview of ecology which has been described. Confucius also accepted the Tao, but focused on the order of human society. Just an nature is an orderly and harmonious realm so ought human society to be equally orderly and harmonious. Confucianism supports an anthropocentric environmental ethic. Environmental destruction, degradation and defilement would in most cases impose deleterious effects on other people and thus violate the first two Confucian virtues i.e. regard for others and justice. A third virtue being wisdom. It would also be plainly unwise, because of being imprudent or profligate and hence violates the third virtue i.e. wisdom.The fourth virtue, namely, faithfulness to one's children or one's children's children or one's more remote posterity is also seriously transgressed. The contribution of confusian thought to universal environmental ethic is clear and apparent.

During the past 15 years of heightened environmental consciousness there has been intense controversy about the environmental attitudes of the Judeo-Christian tradition. Most of this controversy has centred on the relationship between God, Man and Nature in the book of Genesis in the Bible.

There are two separate creation myths is Genesis: (1) The account which begins (rather than ends) with the creation of Man and the Garden of Eden in a single day and is centuries older than (2) the account which begins with the creation of light and division of waters and on the first day and ends with the creation of man on the sixth. The order, even more ambiguous, myth also is subject to two conflicting interpretations about the proper role of Man in relation to Nature. It is in this myth that, one finds that the role assigned to man by God is to dress Garden of Eden (which might be interpreted to mean Nature as a whole) and keep it. This injunction together with Man's naming the animals and thus establishing a kind of power over them and prerogative respecting them suggests the kind of responsible, benign vice-regency of stewardship interpretation.

There are, as well, three possible environmental ethics consistent with the Judeo-Christian worldview, depending upon its interpretation: (1) an indirect anthropocentric, utilitarian environmental ethic associated with mastery; (2) a more direct biocentric environmental ethic associated with stewardship; and (3) a direct biocentric environmental ethich associated with citizenship. While both the environmental ethics associated with stewardship and citizenship are direct and biocentric, they differ in their practical implications. The former would permit benign management of Nature and wise use while the latter would imply a laissez-faire, live-and-let-live approach, incompatible with the present more positive attitude toward environmental protection and improvement. The environmental ethic associated with stewardship is thus both the most practical and the most acceptable interpretation consistent with the Judeo-Christian tradition. Further, since it is a possible interpretation of the role intended for Man by God in both the creation myths of Genesis, it seems the most plausible interpretation of the overall gist of the text as it has come down to us, and its most effective contribution to a universal environmental ethic.

Indeed, current teaching on the environment—as exemplified by Pope John Paul II's Encyclical on the Environment (1990)—stresses humanity's stewardship of nature. People are the guardians, the protectors, of the environment, not its owners. A way of loving one's fellow human beings as oneself, the Encyclical states, is to protect the environment and natural resources on which they depend.

Although the culture and civilization of the Middle East and North Africa are rooted in the Judeo-Christian and Greco-Roman traditions, there is a third major element which spread world-wide, Islam. During the European Dark Age, i.e. Greek science was preserved and developed by Islamic scholars, and Mohammad, the Prophet of Islam, regarded himself to be a prophet of the same God and in the same prophetic tradition as Jesus, Moses and Abraham before him. The Islamic cultural tradition, therefore, has been substantially influenced by Judeo-Christian and Greco-Roman ideas, although it constitutes a distinctive historical and cultural context for environmental ethics.

The Koran is less ambiguous than Genesis about the relationship of human beings to nature. It makes explicit certain themes which are only suggested implicit in the more ancient account in Genesis. According to the Koran, Allah created the first man and woman, Adam and his wife, from a clot, or clay, or dust, and breathed into his creation the breath of life. All other things are explicitly created by Allah for the sake of, the use of, and the benefit of man. Adam and his seed are explicitly made to be the viceroy of God on Earth. According to Islam, then, man is at the moral centre of creation and is, indeed, the very purpose of the creation. As in Genesis, so also in the Koran, it is man's right to have dominion over and to subdue the Earth and all its non-human denizens. Indeed, in the Koran, not only are animals and plants subjected to man, the rivers, the sea, even the sun and the moon are subjected to and subservient to man. Man's dominion over the earth and the subordination of the creation to man is spelled out in no uncertain terms. The Islamic tradition clearly supports, perhaps even more unequvivocally than the Judeo-Christian tradition, a direct biocentric environmental ethic of the stewardship type. The environment, though given over the man's role and subservient to man, is the direct object of respect and care, because it is the handiwork of God and a sign of his power and majesty. The Islamic tradition also clearly supports an indirect anthropocentric environmental ethic.

According to Islam, all human beings are descended form Adam and Eve. Hence all human beings, regardless of race, colour, or national origin, are equally members of one extended family; no people are privileged or chosen; no one is inherently better than anyone else. In Islam, moreover, there is a strong emphasis on justice. Justice, indeed, is one of the cornerstones of the Moslem region. Since environmental abuse and/or destruction are, more often than not, harmful to people they are a form of injustice. To ruin or destroy the environment is tantamount to either bodily injury or the destruction or theft of property or both. Further, ignorance of the complex or delayed effects of action in the environmental arena is no excuse, since Islam stresses the moral importance of knowledge, no less than of justice. These are all truly elements of an environmental ethic with universal implications.

Reviewing the foregoing traditional cultures and religions to find what they have in common with regard to humanity's relationship and responsibility vis-a-vis the environment—in other words, the common ingredients of a universal environmental ethic—is the historic role of environmental education. An ethical attitude toward the environment personally and professionally, individually and collectively, and universally valid, is both the assumption and the goal of this new great field of education, making EE the principle, indispensable instrument for its development.

Hinduism Respects Nature

B M Shukla

The Semitic religions emphasize the utilization of resources of this beautiful earth for the happiness of mankind. All that the earth has is earmarked for man. The emphasis is on use and not on sharing with other life forms; not on reverence and co-existence. Semitic religions took birth in adverse and difficult environmental conditions. How could they lay emphasis on reverence for nature? But they too have warned man to desist from the exploitation of nature and to use natural resources prudently. As against this, in Eastern religions and philosophies, the significance of ecobalance is far more recognized than in any other major religion of the world.

The ancient Indian literature is full of obeisance to nature; it was explained by previous speakers. Ozone layer depletion was foreseen by our *rishis* and that is why when they sang the song of *Shanti* (peace), they also said "Om, Peace to the Ether". Is there anything like that in any Semitic scripture or religion or culture? The idea is not to denigrade any religion or culture. One culture has excelled in one dimension of life while the others in other dimensions. And all these dimensions are important for sustainable progress. So let us take the best and relevant from each culture and put them together to create a universal culture imbued with the wonderful concept of *Vasudhaiv Kutumbakam*.

Another point on which I would like to dwell is the Gandhian philosophy and the development model emerging from it. It was Prof Dikshit who said that this model is no longer relevant and it is only of historical importance. I have respect for the privilege of Prof Dikshit to have his opinion, but let that opinion not go unchallenged because it would be unethical on the part of those who do not agree with his views to remain silent. Gandhi was totally opposed to the silent majority. The majority which does not speak out on issues which concern society, in his opinion, was dead.

The problem of environmental pollution, if we go back 40-50 years, was never heard of except from our scriptures and ancient treatises. All of a sudden, as if we were awakened to the problems posed by the environmental pollution and ecological imbalance, we started talking about these issues. In fact, the common people of India were never asleep. It is the educated man who had a blind faith in the myth of technology, who was asleep and needed awakening. Hundreds of symposia, national and international, have been held to analyse the environmental problems, their origin, causes and consequences. Solutions that they sought in the resolution of the problems was in science and technology, the two major forces which were responsible for the genesis of the problems.

True, science and technology can offer solutions to some problems, and we must use them to our best advantage. But we have almost reached a dead end. There is eco-politics that has halted progress at the international as well as national levels. Limits to what national governments and international agencies can do are already visible.

Village Biospheres of Rajasthan

G S Nathawat

All religious are based on ethical values. Many of them are very old. When they originated, many of the problems like population growth, industrial pollution, etc., were not there, yet they had a strong environmental component.

The unfortunate thing about the religions today is that their custodians are doing everything minus educating about ethics, general or environmental. Even when we have created an imbalance in the environment, we are unable to capture that old religious teaching of oneness of nature. The earth can support only a limited biomass. We, the humans, due to increase in population, have usurped the biomass which which was being used by other living creatures also. That is why so many of the living species have just disappeared from the earth's surface and many more are endangered. It is estimated that by the year AD 2000 we would be exterminating one species every 15 minutes. We will thus destroy the biodiversity.

Biologically, the micro-organisms are more important than the bigger organisms. Micro-organisms multiply at a very high rate. For example, the bacterium divides itself into two within 20 seconds. It does not dies. A single bacterium multiplies into 15,000 within five hours. Any discharge of waste is thus decomposed by bacterium very fast. There are no bacteria in nature to decompose plastics and nuclear wastes.

Then what should we do? Firstly, we should not generate waste which cannot be decomposed by natural processes or nay man-made mechanism. Secondly, we should look for practices in traditional cultures for answers to some of the problems we face today. Ritualistic differences among various religions make it difficult to use them as vehicles of ethical change.

Every settlement in Rajasthan has a common land called *gochar*. In the upper reaches of the watersheds, some land is left out for trees in the name of gods, goddesses or saints. I have studied three districts—Nagore, Church and Ganganagar. I found that much of the land in Rajasthan is overexploited leading to deforestation and soil erosion. But these forests known as *orans* or *devans* stand intact. Not a tree has been touched because of the socio-religious sanctions against cutting the *orans*. They have now become small biospheres of these villages.

On the other side, the forest cover in Rajasthan is reduced to 1.9 per cent even though the forest land, according to land records, is 11 per cent. It is interesting to note that while the forest cover has secularly declined in Rajasthan, the forest bureaucracy and technocracy has increased at an alarming rate. There used to be only one Chief Conservator of Forests in the State; now the number has gone up to 11. There is one Principal Chief Conservator of Forests and a big staff attached to his office.

But look at the *orans*. They have no officer to protect them. Nevertheless, they are safe. It is because of the ethical values and the community-level environmental education of the people.

Atharva Veda and Mother Earth

Sudesh Narang

I would like to say that the Atharva Veda considers all that exists in nature as part of a single soul, *Parmatman*. And thus earth and all that it contains—soil, air, water, animal life, plants etc.—are interrelated.

It covers three environmental dimensions: (1) depolluting the environment to sustain life on earth; (2) bringing peace in the universe; and (3) using prayer as a means to improve the welfare of society. Every element of nature is considered to be a source of strength. We cannot afford to pollute it. We must protect the nature to protect ourselves.

Yagya purifies the environment, but it also means sacrifice, to give and share. The *Prithivi Sukta* or the *Atharva Veda* is all praise for Mother Earth. It tells about the earth and its unity. It is a prayer to the earth to forgive, if man has done any harm to her.

Islam is Pro-nature

Jafar Raza

It has been said that the Semitic religions are not sensitive to nature. No religion negates nature. All religions, Western or Eastern, have respect for nature and consider human beings as integral elements of nature.

If we look at the Semitic religions, we find that their leading figures were either farmers or shepherds. In non-semitic religions, one gets kings, and learned men, etc. To think that a shepherd or a farmer would not have love and respect for nature is beyond my imagination. They cannot go out of nature.

It is true that the Semitic religious have no provisions for worship of nature. The whole *Brahmand* was created and managed by one God. Only that one God presides over it. According to the *Qoran* God sent man as His heir. The inheritor has certain responsibilities. God's creation cannot be destroyed by his heirs, the human beings. They were given the right to use the nature on certain conditions. They were not given the right to destroy nature or to spoil it.

Trees, flowers, fruits, etc., are to be used but within limits. Imam Jafar Sadiq says that before you cut a tree you must plant a sapling at the same place. You cannot cut before planting. Do not pluck the leaves, says Islam, because they worship God. Do not disturb anything. You can use animals as food, but not in such a way that they disappear from the earth because of overexploitation. Animal life is as important as you (human life).

Islam calls for *pooja* or *namaz* five times a day. Before the *namaz*, *wazu* or cleaning hand and face with water is essential. It is obligatory for all. But, according to Islam, if you have so little water that you can save a dog dying of thirst, then you give the water to the dog and do not perform *namaz*, or perform it without *wazu*.

In the *Qoran*, there is the story of Toofan-e-Nooh in which the protection of pairs of various species of animals is mentioned. Even in the story of *Swarg*, or *Zinnat*, there is mention of taba (tree), rivers, water, air, etc. It is true, Islam permits use of nature, but only within limits.

More than 80 per cent of the *Qoran* deals with nature, the sky, the stars, the rivers, the air, the water, etc. It says in clear terms that nature will revenge, if you try to harm it. Islam is *paryay* (synonym) of nature, and is not different from other religions in this respect.

Just saying that Allah is one and Mohammad is his messenger does not make anyone a Muslim. If he accepts Mohammad, he has also to accept 124,000 *Nabis* and *Paigambars* sent by Allah earlier. The *Qoran* says that God sent *Nabis* to every piece of this earth. If you negate one *Nabi*, you negate the *Qoran*. You are not a Muslim. You have to accept all of them. If Allah has sent *Nabis* everywhere, India is included. And hence, a true Muslim will have the same respect for Ram, Krishan, Buddha, and various Rishis and Munis of India as for Mohammad.

We have abused religion to destroy nature. Once Firaq Gorakhpuri said "if I want to see Islam, I see the list of criminals". He used to say such things to provoke others. To provoke him in turn, I said, "the water of my sewage is cleaner than the water of the Ganga which you treat as sacred". He got angry. He said, "I will show you at Hardwar how pure is the water of the Ganga. Here people of cities have made it filthy". I replied to say that the same applies to Islam. Islam in its original form is like the Ganga at Hardwar. A lot of dirt has been added during the last 1,400 years. So if there are practices now which are environmentally unsound, it is not because of Islam. It is in spite of Islam.

Religion has to be seen in terms of its essence and underlying philosophies. This applies to all religions. Let us study a religion before we comment on it. And if we do so, then we will find that all religions inoluding Islam are pro-not anti-nature.

Christianity Believes in Harmony

Father Anto

I would like to mention here that Christianity is essentially oriental in its origin. I do not know, why it is called an occidental religion. Like Buddhism, it has gone out of its place of origin. But the fundamental values contained therein are not occidental. Secondly, Christianity has been on the Indian soil for about 2000 years. It has taken a lot from Indian culture and religions and given a lot to them. I am unable to understand as to why Christianity as practised in Europe or America is taken as a universal phenomenon.

If we are to think of environmental ethics as contained in Christianity, then which Christianity we are taking into account. West has interpreted Christianity as a religion which sanctions control of man over nature. But in India, Christians do not believe in this interpretation.

According to Christian scriptures, God created everything. He is supreme. Human beings in the Book of Genesis have been called the pride of creation. But man has been created with other creatures. He can use nature, the living as well as non-living, but only with *vivek* (discrimination and wisdom). Man has not been given the right to destroy nature. It is in the preservation of nature that lies the welfare of man. Adam named every animal, bird and plant. It does not, however, mean that he got mastery over them. I plead that the Christian West should not be equated with Christianity. This also means that the Christian West cannot talk of Christian values. In India, we after give human names to animals. It is a sign of love and co-existence. It is not a sign of mastery.

We have to make a distinction between good and evil using the power of discrimination. People know that smoking is bad but they still smoke. We have to develop that discriminating attitude.

Jesus was born in very lowly surroundings: in the midst of shepherds and animals. He did not teach the people on the river bank, nor did he live in the mountains. But he spent a lot of time meditating in natural surroundings. Unfortunately, his disciples have forgotten the love Jesus had for nature and all that it contains.

Had Jesus been alive, he would have joined issue on environment. I say this because he took up issues which harmed human dignity and human life. Environmental destruction has serious implications for human survival and development. He would have, therefore, joined issue with us and fought on the side of nature. To sum up, the Christian ethics has its source in the following principles:

(1) God is the creater of this universe and human beings are only trustees to manage what He has created;
(2) Human beings are intimately related to nature. Nature is marked by harmony. This harmony should not be tampered with; and
(3) Human beings should use *vivek* (wisdom) in dealing with nature and environment. They should learn from the life of Jesus and from nature itself.

7

POPULATION GROWTH AND ENVIRONMENTAL ETHICS

7

POPULATION GROWTH AND ENVIRONMENTAL ETHICS

Opening Address

R P Misra

Population, i.e. the quantity and quality of people, is a dynamic force. It is not static. It changes, slowly or rapidly, depending on a variety of factors. Population by itself would have had only an academic value but for the fact that the nature which provides it the wherewithal to survive and prosper is not so malleable. It is dynamic only in the long run of millenia. People's welfare depends, to a great extent, on the man-land ratio (where man=population, and land=nature).

Man interacts with nature through the medium of culture. Culture provides the values, knowledge and technology for man-nature interaction. So long as the number of people interacting with nature is within bearable limits, everything works well. But when the number increases fast and the man-land ratio becomes adverse, as in much of Asia, Africa and America today, ecological and environmental problems emerge. There is another situation in which these problems become serious. It is when cultural values become insensitive to ecology, as they are now in the West.

It so happens that at present both the forces are at work. The Western world is consuming so much and wastings so much that nature is unable to meet the demands put against it. The less developed and poor countries are growing so fast populationwise that despite their low level of consumption, their total demand on nature is growing too fast for the known resource base. The environmental and ecological problems faced by the world today are partly rooted in the rapid population growth in the South and partly in the consumerist culture of the North. We discussed the cultural dimensions in the previous sections. We will examine the implications of population size and growth on ecology and environment in this section.

Around AD 1650 the world population is estimated to have been half a billion. The most populated continent was Asia containing almost three-fifths of the total population of the world. By 1990 Asia's population had tripled while that the Europe quadrupled. And if we include the major chunk of the American population of European origin, Europe's population increased five-fold.

ESTIMATES OF WORLD POPULATION
1750-2000 (in millions)

	1750	1800	1850	1900	1950	1970	2000
World	791	978	1262	1650	2515	3535	6130
Developing Countries	590	731	919	1088	1682	2580	4742
Asia (ex. Japan, USSR)	468	600	770	881	1298	1953	3336
Africa	106	107	111	133	222	344	768
Latin America	16	24	38	74	162	283	638
Developed Countries	201	247	343	562	834	1055	1388
Europe (ex. USSR)	125	152	208	296	392	462	527
USSR	42	56	76	134	180	243	353
North America	2	7	26	82	166	228	354
Japan	30	30	31	44	83	103	122
Oceania	2	2	2	6	13	19	32

Source: J D Durand, *The Modern Expansion of World Population, Proc. American Philosophical Society*, 111, 137 (1967).

In 1650 Asia accounted for 60.5 per cent of the world population. It came down to 58.25 per cent in 1900. The share of Africa came down from 18.35 to 7.46 per cent while that of Europe increased from 18.35 to 24.95 per cent, and that of Americas from 2.40 to 9 per cent.

The West was the major contributor to population growth until 1950. It was also practically the sole user of much of the earth's resources and sole polluter of the environment. Only after the World War II, the population of Africa and Asia increased at an accelerated rate while that of Europe slowed down. But the per capita consumption of energy, food, and industrial goods continued to rise in the West offsetting any gains from the slower growth of population.

The rapid population growth in the developing countries since the 1930s has only added to the environmental problems already set in motion by the West. Increase in population has an spiralling effect on demand for natural resources. And the adoption of the Western style of development with emphasis on mass production and consumption has further accentuated the pressure of population on land.

The most serious environmental implication of rapid population growth in the developing countries is deepening poverty of the people. Man-land ratio has become adverse. In India per capita availability of arable land has come down to 0.3 acres. When we add to this the inequities in distribution of land and other natural resources, the situation appears really serious.

Lots of people are living a precarious life. They toil for just two square meals. They are shelterless. They have no earning capacity. Many of them drift to the slums and shanty towns of cities. Half the urban population of India lives in these informal settlements. The cities are bursting at the seams; they are thoroughly polluted by industrial and domestic waste. Clean air and water are no longer free gifts of nature; and access to land is scarce for all and impossible for the poor majority. The quality of life in urban India is really depressing. Human drudgery, deprivation and indignity constitute the worst aspects of environment. It is the greatest pollution one should be concerned with.

The situation in rural India is not far better, if not worse. The village are also overpopulated in relation to resources. Educational level is low, the real literacy rate being less than 30 per cent. Those who get education drift to urban slums and industrial jobs. New technology needs specialized and few people to man machines. Slums too are overpopulated. Outlets for jobs outside the village are drying up. Rural economy is growing too slowly to generate non-farm jobs in sufficient quantity. The unemployed youth has nothing to do. He is not wanted in agriculture not is he wanted in urban industry. He lands himself in criminal dens and finds temporary relief in drugs of various kinds. Village environment is now highly polluted. It is no longer peaceful and serene.

Population pressure is having telling effects on village ecology. The forests, the trees, the pastures, the tanks and other village commons are now gone. Natural greenery, arable land, potable water, and clean air can no longer be taken as given.

Enough is not being done to check population growth. The West which overpopulated the world to being with is no longer interested in the problem. It sees this problem from a distance and takes a pseudo-moralistic view. The developing countries, in this last quarter of the 20th century, cannot force on people anything, howsoever desirable it may be. Its political fallouts are often devastating for the ruling classes. So, in spite of lots of money spent on family planning, all it has achieved is the awareness about the problem among the educated. Much of the illiterate and poverty-stricken people go about their business as usual. Unfortunately they constitute a majority in most developing countries.

The close relationship between poverty and population growth is only now being taken seriously. A large number of children is a part of survival strategy of the poor. So long as survival is not assured, the prospects of lessening population fast enough to give expected results are dim and thus the prospects of an ecologically balanced and environmentally clean society are also dim.

This calls for a truly global effort. Every society has contributed to population growth at one time or the other in the recent past. All have equal responsibilities. Developing countries along cannot be blamed for what they are doing now; others have done the same earlier. Poverty eradication and environmental protection should constitute the prime agenda of the United Nations. They must go together.

Science and technology have made it possible to reduce the death rate. It is now the task of ethics to bring down the birth rate. This calls for a new global ethics reasserting the unity of life, unity of humanity and unity of the Earth. It calls for a reproductive ethics which promotes sustainable development.

Population-Poverty Nexus

Ashish Bose

Population growth in India in the post-World War II period has acquired serious proportions, and our inability to slow it down has put the economy and ecology of the country in jeopardy. The whole subject of population, environment and resources is a highly complex question. I just read the 10-point programme of UNFPA on population and resource development. There is nothing to disagree on what it says, but then it says so little and so generally that I cannot see any outcome except public awareness.

In a country like India, we cannot take a macro-view of the environment. We have a great variety in geography: deserts of Rajasthan, mountains like the Himalayas, river valleys, the sea coasts, forests, plateaux, etc. When we are talking of environmental ethics in the context of increasing population, we have to ask the question, which environment?

I would like to add that population studies cannot be done by lone scholars. We need teamwork. No economist, geographer or demographer, howsoever qualified he might be, can do this work singly.

Now, let me give examples of population growth ethics as practised by the people. I was doing field work in Rajasthan (Churu Distt) during drought times. Our general impression is that rural people like to have sons but we do not know, why? When we asked one of the respondents the number of sons he would like to have. He replied "two". When we found that almost every respondent was giving the same reply, we cross-examined to find the reasons. When we asked some women as to why they wanted two sons, they said that my question was a foolish one. They put a counter-question to me. Why do you have two eyes?" I could not answer. And they answered for me. Suppose, one of your eyes is gone, then you have at least one eye to live on.

This is the social behaviour; you may call it demographic fundamentalism. You cannot question this. This fundamentalism, however, has its rational basis. People had to get drinking water from two kilometres or more. You cannot send small girls to such long distances. You need boys who can walk or ride on camelback. So the desire to have two sons has something to do with the strategy of survival.

Mohammad Abdul Mannan, currently with Bangladesh Institute of Development Studies, found in his doctoral study that people wanted sons because they were needed during storms and floods. You must have sons to overcome such natural disasters which visit Bangladesh so frequently and ferociously. The chances of survival of households without sons are dim. The attitude towards sons is thus rooted in the strategy of survival. And it so happens that many Bangladeshis in search of sons end up with five to seven daughters.

Now, let me give you an example of governmental ethics. During the time of drought the people working in Churu district were given Rs 11 as wages. Most of them were women. The minimum wages fixed by the government was Rs 17. We found that no one was being given the minimum wages. The government functionaries said they were permitted to give less wages during the drought so that more people could be employed. Bunkar Roy, a well-known activist social worker, took up the cause of the people and had to go to the Supreme Court to get the statutory minimum wages restored. Even today, the minimum wages law is not being implemented everywhere in the country. What ethics is this? Even the government exploits the people in distress.

The only way to resolve such problems in countries like India is case studies, scientific investigation and teamwork, on the one hand, and a far greater involvement of social workers and NGOs than we see today, on the other.

Perception of the Poor is Different

H N Misra

While I do feel that the environmental problems have become acute, I don't think the humanity has reached a dead end. Human civilization has weathered many a crisis in the past. The present crisis will also get resolved, although it may leave scars deep enough to be visible for decades and centuries to come.

Man-nature relationship has taken three forms: (1) man in harmony with nature, (2) man as subordinate to nature, and (3) man as the master. All religions have emphasized the first one. The second one was branded as environmental determinism and rejected in the early part of this century. And the last, i.e. man as the master, has been the guiding theme of industrialism which grew out of the European Renaissance.

There has always been a conflict between conservation and consumption; and between preservation and destruction. Very few societies in modern times have been able to resolve these conflicts satisfactorily and strike a balance. It is for this reason that environmental issues have come to the forefront.

The man-nature relationship is largely determined by the ethical values. Formal laws and scientific knowledge, unless backed by ethics, can easily distort this relationship. I am, therefore, glad that this symposium has been organized to discuss the ethical dimensions of the environmental problem. Environment is the concern of the people and to understand how would the people look at their environment, we must know their ethical values.

Some of the colleagues who have preceded me have talked about the rapid increase in population as a major factor upsetting a harmonious man-nature relationship. I fully share this view, but the population growth apart, it is the concentration of population in large cities which is the major culprit. If rural development is promoted, migration to cities will decline, and the urban environmental problems to that extent will be lessened.

Our society is divided into the poor and the rich. The rich perceive the developmental problems differently from the poor. The basic needs of the poor have to be met, no matter what happens to the environment. And if environment is the priority of the rich, then they must reduce their conspicuous consumption at least marginally, so that the living conditions of the poor are improved without putting any additional stress on the environment. It is imperative that the parameters of environmental ethics take into account the perception of the poor while evolving the new code of conduct for interaction between man and nature.

Case for Family Planning

Rajiv K. Sinha

Nature is so constituted that there are always more producers than consumers. Look at the flora the world over. The plants produce so much for animals and humans. But what happens to plants when the swarms of locusts invade. Plants come to life soon after the swarms disappear from the scene. What has happened during the last hundred years is that the consumers have overtaken the producers. Man armed with the so-called modern technology has destroyed plant life and created an artificial demand for goods not necessary for a healthy and cultured living. The very base of the stable pyramid of life is being degraded through clearing of forests. Anything that we do to destroy this balance in nature will affect someone adversely. It may be the person who is indulging in this exercise along with his innocent fellow beings. This is a patently unethical behaviour.

Let us take a lesson from biology. The population of yeast increases very fast but it finally declines and stops. It follows a J curve. We should foresee the dangers ahead and correct the situation.

D N Chaturvedi

Let me emphasize a neglected aspect of population. In countries like India, children are not only an economic necessity but also a cultural necessity. While it is good to control population let it not lead to a negative population growth and promiscuity. What would be the impact of various birth control pills and methods on the child-bearing capacities of women, we do not know. But its impact on morality is clear and visible to us. The promiscuity is leading to AIDS and other diseases. It is my plea that population planning should be synonymous with social and cultural planning so that self-control becomes the prime means of birth control.

So long as we rely on technology even for social development and treat human beings as machines, we will have unsuspected side effects. Sex is an issue that involves emotions, hopes, desires and intimate human relations. It is not a factory to produce children, as many of us tend to think, which can be opened and closed down by the government.

I am all for population control but not the way our population experts prescribe. I am glad that Prof Ashish Bose asked for an integrated approach but somehow he also fell in the mechanical trap, and did not take full note of the cultural and human dimensions of population planning.

Environmental Implications of Urbanization

J R Shukla

I would like to stress on a problem which has by and large been neglected by the Symposium and which is the direct outcome of the development model we have pursued so far. That is, urbanization. High level of urbanization is considered synonymous with high level of socio-economic development.

It is obvious that urbanization has created a number of serious problems and some of the major environmental and ecological problems we face today have emerged from the urbanization-industrialization process. That is why the problems of air pollution, water pollution, overcrowding and slum-dwelling, energy crisis, etc., are seen more in cities than in rural areas.

Urbanization of modern times is the result of migration of rural people to urban centres. And this migration is caused by relatively more employment opportunities in urban centres (pull factor) and diminishing opportunities in agricultural villages (push factor). In the context of increasing population, migration is the means to redistribute people in accordance with employment opportunities.

This has given rise to a series of problems. Firstly, it has destroyed the village community and its agro-industrial base. The rural areas have been drained of quality population. Urban industries have practically destroyed the traditional rural industries. At the same time it has given rise to a large number of problems in cities where one quarter of the population of India lives.

The continuing rural-urban migration and high rate of natural increase in urban population has created a situation where there are far more people entering the labour market than the jobs available. Cities thus not only face the problems of overcrowding, housing shortage and consequent growth of slums but also problems of unemployment and poverty. More than one-third of the urban population of India is below the official poverty line. In some of the states like U.P. and Bihar, this share touches 50 per cent mark. Thus by 1991, about 100 million urban dwellers were poor. Such a situation has made urban communities unstable in India.

So one of the major contributions of modern Western-style development, which Prof Dikshit says cannot be avoided, is to create problems that defy solution. This calls for a reappraisal of the current urbanization process. As suggested by Prof Misra, India should adopt a futuristic culture-rooted urbanization avoiding too large cities and promoting small-and medium-size towns.

What ethical values we can expect from a society which cannot given even a place to live, or which forces 10 people to live in one room; which has no employment opportunities and throws the labour force to fend for itself in the informal sector of the economy. And what environmental quality we can expect in a situation where majority of the people have no stake in the city. And added to this is a fairly large chunk of population which overuses the available resources and releases pollutants which make life difficult for others.

K D Gangrade

I would like to point out that the urbanization process triggered off by the Industrial Revolution has created more problems than it has solved. Eight per cent of the cities of America are suffering from pollution which causes cancer and other diseases. Since 1975, they are spending huge sums of money annually to solve this problem. We are getting into the same trap. Most of our cities are highly polluted. Unless our style of life changes and we develop a social order which takes care of nature, such problems will continue to grow in number and intensity. Gandhi gave a holistic view of life. He also said that we should learn to live with nature. We should now try to answer the following three questions to bring about a new social order:

1. What are the limits to economic growth? Wants are insatiable but the resources are limited.
2. How to bring in the ecological criteria in our developmental programme? Economic cost will go up if environmental costs are incorporated in project costs. What social cost we are prepared to pay?
3. What are the specific goals which we should set for technological development?

8

SCIENCE, TECHNOLOGY AND ENVIRONMENTAL ETHICS

8

SCIENCE, TECHNOLOGY AND ENVIRONMENTAL ETHICS

8

SCIENCE, TECHNOLOGY AND ENVIRONMENTAL ETHICS

Opening Address

R P Misra

Earlier chapters have dwelt on this aspect. Here a few points not mentioned and emphasized there need attention. It is unfair and unreasonable to blame science and technology for the environmental ills of the modern world. Science and technology, no matter what their origin, should be treated as great achievements of humankind.

The problem does not lie in science and technology; it lies in their speed, sweep and comprehensiveness and in the inability of the philosophical and humanistic pursuits to match them and to circumscribe their transient use by ethics. Thomas Traherne, the 7th century mystic, writes movingly,"He that knows the secrets of Natures with Albertus Magnus, or the motions of the heavens with Galen, or the cosmography of the moon with Helvetius, or the body of man with Galm or the nature of disease with Hippocrates, or the harmonies in melody with Orpheus, or the poetry with Homer, or the grammar with Lily, or whatever else with the greatest artist; he is nothing if he merel knows them merely for talk or idle speculation or transient and external use. But he that knows them on his own, shall profit infinitely".

Science and technology, and philosophy and ethics, must be balanced. Either of them without the other goes haywire. The ascendancy of science and technology downgraded philosophy and ethics. The rationalist even tried to find a place for God in his picture of the world. But God whose centre is everywhere and his circumference nowhere could not be fitted into a diagram. He is the convas on which the picture is painted or the frame in which it is set, says late Dean Inge.

The new communication technology has the power to mesmerize the minds of the people. When used in large doses, it works as a sedative. It gives a feelings of bliss and conjures visions. By the use of systematic violence, it brings individuals to a stage of utter fatigue and makes them highly suggestible. A new man is recreated in the image of Techno-God. These new men and women are no longer interested in freedom and dissent. They are interested in whatever they are told to do. Individual virtues no longer count.

So, it is the failure of the humanistic disciplines to stand up against the scientific and technological onslaught intentional or unintentional, direct or indirect on man and human spirit. To change the situation, attack on science and technology is not enough. It will not take us far. The most important task is the strengthening of the freedom of individual within the bounds of a humanistic code of behaviour.

To ask an illiterate man not to cut a tree because is gives oxygen and has aesthetic values will not help. Even a formally educated person does not think that his cutting of a tree is a serious below to environment. But it does make a difference when a tree is seen as an element of the cosmic life. This latter belief protected the environment in the past. Natural phenomenon symbolized gods, goddesses, sisters, brothers, friends, etc. Nature was humanized.

With the new scientific knowledge, many of these beliefs have proved rational once again. The Bodhi tree under which Lord Buddha achieved Enlightenment is sacred in India. Many Hindu gods reside in and on it. The recent discovery that this tree gives oxygen for more than 20 hours a day has revealed even to the rationalist why Hindus and Budhists consider this tree sacred and water it daily.

There is a scientific basis for much of the rituals and beliefs of each and every great tradition. Intuitive bases for many scientific discoveries cannot be totally ruled out. Science and religion must go together. Then alone the two could serve the society well. Here lies the basis for new Environmental Ethics.

Technology and Environment

B R Nanda

Ethics, in my view had nothing to do with scientific discoveries and inventions. It was the curiosity in the mind which led to the development of science. The scientific ideas were applied in the 18th and 19th centuries for the development of industry, trade, communication, railways, steamship, mass production, etc., that changed the face of Europe.

With such powers, Europe conquered the Africas, Americas and much of the rest of the world. That process continues even today in one form or the other. Science was used in developing a capitalist economy.

People took to machine manufacturing because it needed few people and it was more profitable.

Even today when science becomes technology through R&D, they want to see whether it is cost effective. Many of the discoveries of science are left alone until the time comes for their commercial application. The penicillin was invented long before the World War II. But the incentive to make it commercially viable and to it within the range of common use came in World War II. Many people were wounded and died, so the need for a drug like penicillin became all the more urgent. The same applies to atomic energy. It is World War II which hastened its development.

Unfortunately, scientific research in the universities requires government funding. Scientific research is a search for truth. But the use of the truth depends on who supports and pays for it. When the atom bomb was first exploded, Oppenheimer, who was closely associated with the project, quoted from the *Geeta* where Lord Krishna says, "I am the creater and also the destroyer." What had been created was terrible, that is how the father of the atom bomb felt. Later on he was associated with the hydrogen bomb also built he felt guilty for having been a party to such weapons which could destroy so much life on the earth. But I think most scientists are not bothered like Oppenheimer. A scientist takes a great risk. But that is not the way he thinks.

Jawaharlal Nehru was thoroughly convinced that without the application of science and technology, a country like India could never get out of backwardness and poverty. He also was convinced that scientists were a sort of people who were in search of truth: not interested in anything else. Scientists, like anybody else, are not generally concerned with the ultimate results, they are concerned with what happens in the small framework in which they work.

I wonder if there is enough teaching of humanities in our scientific and technological institutions. I am a student of history. As far as my memory goes, the humanities do give us the opportunity to think about the problems of humanity: what the Greeks thought; what the Hindu and Buddhist sages thought, etc. For example, it is just not possible to sell Gandhi to a scientist or a technologist. It is easy to sell him to a historian. The kind of mind a scientist is trained to have is not interested in ethical questions.

In essence, the basic idea of technology is to have control over nature. The basic idea of ancient thought was to work harmoniously with nature. This aim of control over nature has done great harm during the last few decades. During World War II, vast resources of Germany, America and other countries were used for certain ends. Star wars, atomic warheads, etc., need billions of dollars and thousands of scientists to work on projects that promote them. This is the direct consequence of investment in science and technology which produced arms and armaments. But then, when I make these remarks, I am conscious that we cannot stop our R&D work only because it has done great harm in the past. The problem really lies with our inability to evolve countervailing ethical values to control the technology going astray. What I suggested is a sort of temporary slowdown or halt in some of these activities and a massive movement to recreate ethical values.

Let us take the case of Punjab. The chemical fertilizers have raised the levels of production of wheat and other crops. The Government of India has earmarked Rs 5,000 crores as subsidy on fertilizers in 1992-93. But Borlaug who is the father of the Green Revolution now regrets having advised the world to go for chemical fertilizer on such a massive scale. It is exhausting the land. If the chemical fertilizer input is reduced to half in Punjab and an equal amount of organic manure is given to the land, the productivity levels would not decline. This has been demonstrated there. This abuse of the scientific discovery need not be reversed but it certainly can be tempered.

We all know that forest degradation has been going on for quite a long time. The forest contractors, bureaucrats and politicians collude to destroy the forests. We allocate more money for afforestation and forest protection. But what we have are less and less forests and more and more of foresters. No one is questioning the right to use resources; we are only questioning their uncontrolled misuse.

Gandhiji said that there is enough in this world to meet everyone's need but not everyone's greed. We can say that there is enough for the need of every nation but not enough for the greed of even one nation. The assumption that Gandhiji made was that man was essentially a spiritual being. He would get his fulfilment not only in his physical needs but also in something beyond it. Here his ideas conflict with ideasof those who think that the fulfilment lies in "eat, drink and be merry".

Gandhiji once said that there must be a fullstop to human appetite because at a certain stage, it becomes totally selfish and meaningless. This included even intellectual exercise. But the point is, how to impose this discipline. He never imposed any discipline. It must come through self realization. This takes us to education; to a certain way of life which in our country we have seen during the last 30 or 40 years being polluted to the bottom level.

Another thing I would like to say about Gandhiji is that his ideas did not come out of any seminar. His ideas were based on certain moral and ethical foundations: there must be something for everybody. *Sarvodaya* means welfare of all. Welfare of not only the majority, not only of a few countries, but of all: not only India, but of all nations. He was not swayed by or tied to any fixed approach or idea. For example, we sometimes do a great wrong by associating ourselves with *khadi*. Gandhiji promoted *khadi* because it could offer something to those who had nothing; because it promoted self-reliance. It carried the message of freedom. He asked people to wear *khadi* even by paying more for it. But now, should our approach to *khadi* remain the same? I was startled when I read that Gandhiji had once said that the time may come when you may not like to produce cotton where cotton is produced; you might like to grow some other crop which is more remunerative. You may have reasons for which *khadi* cannot provide enough remunerative employment and income. In such cases to say that you go on producing *khadi* will amount to saying that a man should not go round the ancestral well but must drown himself in it. Gandhiji thus does not inhibit us from thinking afresh. Whether it is the problem of environment or organization or politics, he gives us a framework or attitude with a great sense of humanity.

There is, therefore, no inherent antagonism, in my view, between science and technology on the one hand and environment on the other. What is needed is a moral awakening of scientists and scientific tempering of the spiritualists to resolve this conflict. The conflict is not real; it is the product of disjunction between science and spirituality.

Scientific Pursuit and Ethical Values

S K Sinha

The knowledge that we derive from scientific pursuits is not absolute, it is relative. The conclusions that the scientist arrives at are based on the knowledge that exists at a given moment of time. Whether it is motivated or genuine is very hard to say; and very difficult to ascertain instantaneously. When scientists who have acquired stature say something, people believe that what they say is based on conviction and scientific inquiry. But once the issue takes a political dimension as in the case of atomic energy or ecology, scientists are prone to advance biased opinions unless they follow appropriate professional ethics. Way back in 1980, there was a group of 17 experts representing the World Meteorological Organization, UNEP, etc. which met at Vienna in Austria and issued a statement about carbon dioxide. I happend to be one of the 17, only one from the developing countries. There was no talk at that time about other environmental problems. Estimates being made about deforestation around the world were very uncertain. What was certain was the emission of carbon dioxide form energy consumption. And, therefore, the whole emphasis was on energy. As a novice, I put a simple question. What would be the situation if new technologies develop to harness solar energy? In that case the question of emission would not be there, and the energy would be available to all. I was told that it was a political question!

Later, the issues of global warming and ozone layer depletion came to the forefront. Search for causes and remedies started. Just five years back or so, suddenly it was found that the developing countries of Asia, because of a long tradition or rice cultivation, were contributing a lot to the Green House effect. And, therefore, they were equally responsible for polluting the environment. And thus the whole problem took a political rather than scientific turn. We had to wait because the scientific data were not adequate. Now we have the data to counter this patently wrong conclusion. But what we find now is that when we want to publish our findings, it is not readily accepted by the Western media. So, while at the national level, we talk of different aspects of pollution and try to find falut with each other, a similar situation prevails at the international level which must get our attention.

In 1987-88 the G-7 met when Mr Carter was the President. He asked the views of the National Academy of Sciences of the USA on energy in future. They told him that he should not raise this topic in the G-7 meeting. But if he did, the problem that would emerge soon would be a major shift in energy consumption from oil to coal. And at present of the coal is controlled by the USA, USSR and China. USA must make an effort to arrive at an understanding that the three of them will decide how the coal should be utilized by different countries. In other words, how the industry will be run in the world.

Now this is a published document which I will be happy to pass on to anyone interested in reading it. So what I am saying is not something abstract. What is being said in the open is one thing but what actually goes on behind the scene is quite different. The United States is not interested in considering any kind of reduction in CFC, etc. It is only Europe which is very conscious of it. The USA will not do anything which would adversely affect its world supremacy even in the remotest future. But we know that beyond the future of the nations of the world is the future of mankind. To safeguard that future, we need strong ethical values.

There are two persons who have consistently talked about environmental ethics. One is Mrs Gutland who headed the Gutland Commission or Environment and Development and the other is Mr Maurice Strong, Chairman of the Rio Conference. They have been arguing that the developing countries should be given at least 15 years not to slow down the rate of progress to industrialize themselves. On the other hand the developed countries should slow down their industrialization. But the USA would not buy this idea.

The so-called truth of science at any given point of time is not recessarily the truth. It has to be evaluated by others. Secondly, there are people who are concerned about the environment, about human destiny, but it is economics and politics that determine the decisions. Two years back the National Academy of Sciences, USA, suggested to the present President of the USA that in the past international relations were determined by security and economics. But in the next century, they will be determined by environment and ecology. How many concessions can be given and taken is going to be the important issue.

No one will deny that every nation including India should now move towards sustainable systems, but the fact is that when we look at the situation, we see that this country with 850 million people produced only 176 million tonnes of food this year and is forced to import one million tonne. If the population grows at the rate it is projected, and our requirement grows simultaneously, then keeping in view the change in the food hahits resulting from improvements in the per capita income, etc., by the turn of this century, we need about 250 million tonnes of foodgrains. It means that there has to be additional production of the order of 80 million tonnes. Given the resources that we have, it appears to be a difficult proposition. Alternatively, we should have enough foreign exchange to buy food from outside. If neither of the two approaches is practicable, then there will be a real shortage of food for which none of us is prepared.

The countries which have been able to achieve a breakthrough in agriculture such as China, have a far more input of fertilizer, etc., than India. China uses 14 million tonnes of fertilizer. Out of 94 million ha China cultivates, 60 million ha are irrigated. Added to this, the climatic conditions there are similar to those in the rabi season in India. Temperatures remain low for much of the year. And finally, Chinese have successfully used their organic waste.

When we look at our situation, we find so many difficulties ahead. How many people will be ready to use human excreta for compost? In Bombay University when excreta was used to produce cooking gas, students refused to eat food cooked on biogas. The other problems are well known. So let me not repeat them here.

At the international level, discussion are going on for conventions/ treaties on Climatic Change, Deforestation, Biodiversity, etc. Work is already on to determine how the world should regulate itself. It is here that the scientists of this country are called upon to make contributions by appropriate data. Unless we have the database, we will lose our ground in the comity of nations. While ethical issues are important, let them not degenerate into polemics. Let them emerge from sound and reliable information.

Role of Business and Industry in Enviornmental Recovery

S Neelamegham

I am going to talk from the business and industry angle. Business and industry have a major role to play in matters environment. The are the culprits in the public eye and are being critized very harshly. But I would like to say how other countries have gone about in this regard, where are we now, and what are the kinds of steps that we must take to safeguard our own environment. I strongly believe that technology is not bad or good by itself. Much depends upon how we use it. We can use it to improve the quality of life. We can use it otherwise too . For this cannot blame technology. We have to blame ourselves and our institutions.

We have passed through three technological revolutions so far. Two hundred years ago, we invented steam power to run railways, mines and industries. And 100 years ago, we brought about another kind of revolution when we started using electricity. As a result we are today using telephones, elevators and skyscrapers. All this is possible today because of the technological revolution which took place only 100 years ago. Today we find ourselves in the midst of a third revolution centred on computer, telecommunications, etc. We find today voice-telephone, image telephone, fax, and electronic mail system. All these are realities today which we call the third technological revolution.

One good thing that the three technological revolutions have done is that geography as a control variable is now very much weakend. The whole world has become one. Even in Delhi there are very few people who do not have the facility of a dish antenna to watch different kinds of programmes broadcast by other TV stations. You cannot stop it because the communication revolution that is taking place in other parts of the world is known to everybody and everybody wants to know. That is the reality.

But of course when you talk of Gandhian philosophy, then let us keep in view that Gandhiji had a different worldview. He always thought ahead, looked ahead and what he said is far more real. There is no doubt about it. He always emphasized the village as the foundation. In his own words, "if India is to attain true freedom and through India the world also then sooner or later the fact must be realized that people will have to live in village, not in town; in huts, not in palaces". Of course it is very difficult for many of us if not all of us, to practise his ideals. If we ask doctors to go and serve in villages, they don't want to go. You ask the bureaucrats, they don't want to go. But still the truth is that unless and until we bring life to the village, the country will not have a bright future.

Nehru had a different vision. He wanted that the country must become industrially and technologically powerful. That is why we went in for investments in energy projects, in electronics, in industry, in agricultural research, etc. No doubt as a result of these investments, we could develop a vast infrastructure not only in terms of industry but also in nuclear energy, science, technology, etc.

Today we are living in an altogether different era. India has just decided to integrate itself with the global economy. The reality of today is that we cannot live in isolation. All our neighbours have already integrated themselves. But is this the right direction for the world to go? Had Gandhiji been here, he would have said No. And we are reaping the consequences also. But that is how the world goes. We cannot stop it. But the ideals that Gandhiji put before us need attention. I will come to it a bit later.

What has been the industry's response the world over? Technology has good effects as well as bad effects. Industry presents a paradox when we consider the development needs. The inherent process of industry will affect the environment. But certainly the processes can be modified so that one can lessen if not completely eradicate the menace that is destroying the environment. The 1972 Stockholm Conference on environment promoted new thinking. That is the beginning of the idea that the industries be also aware of what they are doing to the environment.

This naturally created a problem for the industry. Industry/business is profit oriented. They never thought of pollution abatement costs. Psychologically they could not even think of spending money on pollution control. They are so much used to investing in projects that increase productivity and profit that they could not think of investing in pollution control.

Of late, during the last 10 years, many changes have, however, taken place even in the case of industry. The UN programme on environment is also there. Today, every product is seen in the light of how it will affect the environment. Putting more chemical fertilizer is not the only way to increase agricultural productivity. Perhaps reducing chemical fertilizers and increasing use of organic manure will be cost effective, both from input and output angles. If we do not make a right combination, it will affect the soil fertility. In fact all over the world there is rethinking about the use of fertilizers and pesticides.

We have identified the causes and solutions for many of our problems. For example, consideration of recycling of water and other wastes had become an integral part of industrial project planning. In Europe, the industries have to complete with the USA and Japan, and yet have to comply with the environmental standards set by the European Community. The Community has become very conscious of environment. You also find that it is normal tendency for anyone including industry to find the least cost method of production. You have to demonstrate to the industry that by following certain kinds of methods you not only lessen damage to the environment but also make more money.

Many industries are today realizing that if they use the antipollution devices, they may increase profit. Recycling can become a profitable proposition. A survey in Holland and Germany and Germany revealed that 60 per cent of the Dutch and 80 per cent of the Germans, when they went to market to buy any product, wanted to know how the product affected the environment. Half of the Britons, a recent survey revealed, wanted to buy environment friendly products.

The International Management Institute at Geneva conducted a survey to top managers. Sixty-seven per cent of them said that they were going to invest large amounts of money in order to prevent pollution. They consider it is good investment. In all European business schools today environmental management is a compulsory subject. Stock brokers in the UK give weightage to those organizations which are environmentally effective. UNEP says that in 17 out of 27 countries of Europe, carbon dioxide emissions have declined; paper mills have reduced water consumption from 180 to 70 cu m per tonne. Coca Cola has started recycleable packaging. Because of the consumer movements in these countries, retailers give preference for those products which are environment friendly. It has become a selling point. In Denmark there is a very big departmental store called *Erna* which has removed all hazardous substances from packaging. *Migros* in Switzerland has done the same.

A few suggestions I give before I conclude. One is that action is needed. As Prof Misra said at the time of the inaugural function, legislation in this country somehow or the other does not work. In Europe it works. In America it works. But in India we do not have dearth of legislation but nothing works. Why? In fact, the NGO activitists are the people who bring results. Governmental machinery just does not work. So a collective action is required.

Secondly, target-setting at the industry level is a must. Instead of troubling the consumer, standards should be enforced at the production level. Emphasis must be on energy conservation and we have to apply existing technology to reduce hazardous emissions. We should also label products so indicàte environment friendliness.

We have to create environment educating programmes to create social awareness. People are not yet aware of the dangers. I strongly believe that we cannot go back, but we can move forward only if we practice the ideals of Gandhiji.

Towards a Dharmic Frame for Business and Industry

M B Athreya

It gives me great pleasure in being here this afternoon on several counts. Universities all over the world in the sixties and seventies were the home of two kinds of things. Either they were dealing in econometric growth economics or in Marxian economics. That tradition continues in many of them even today. I am glad to see that the Gandhian ideologies and Gandhian philosophy have survived, for they alone provide the third alternative and, if I may say so, the only sane alternative for the future.

I spent a year in California in 1964-65 when the free speech movement began. It was a period of great trouble in India too. The Naxal movement was at its height. We went through *gheraos* and violence on the campus. In 1969, in the faculty of IIM, Calcutta, I was in minority of one writing a paper on Gandhian thoughts on industrialization, and social responsibility of industrialists. So, I readily agreed to Prof Misra's invitation to talk on business responsibility and environmental ethics.

I would like to emphasize a few points which are discussed in some detail in my paper. You may recollect the debate on welfare economics.

Milton Friendman has gone on record saying that social responsibility in business is a dangerous concept. Social responsibility is the business of government, politicians, bureaucrats; and it will be very dangerous for a businessman to worry about it. Business is for profit. Galbraith, at the other end, has been a very severe critic of business. In his book *Affluent Society*, he has discussed the damage that business can cause to society. For some time social scientists tried to create value-free social sciences. I am glad that we are now concerned about values in our academic discussions. This is in line with our own tradition which lays a lot of emphasis on values and ethics.

Indian culture seeks four objectives of human existence: (1) *dharma* (2) *artha* (3) *kama* and (4) *moksha*. All the four were considered equally important but only in this hierarchic order. What capitalism and Stalinist communism did was to put *artha* at the top, emphasized *kama*, sought *moksha* in material gain and almost forgot *dharma*. Environmental ethics is a part of *dharma*. We have to start giving serious thought to it because there is urgency about it. It is better late than never and amounts to making virtue of necessity. Environment is a shared responsibility.

Business is an institution that we hate in India for both right and wrong reasons. It has emerged as a panacea for the failure of central planning but it is considered to be insensitive to social responsibilities like environmental cleanlines. I must, however, add that even if we do away with private business, still there will be environmental problems.

I will like to suggest seven different ways business can discharges environmental responsibilities:

1. *Raw materials and inputs:* Mining damages land. The USA has a legislation under which companies cannot declare dividends and profits unless it is certified that environmental damage, if any, caused by them has been repaired. Companies using agricultural raw material can ensure that excessive amounts of fertilizers and pesticides are not used in the production process. Long-term damage to soil can be avoided.
2. *Inward transportation:* Companies can stop transportation of nuclear and other hazardous materials to the factories or at least they can bring them more safely.
3. *Storage:* It can be done much better to minimize the fallout.
4. *Process side:* Much of the environmental legislation has treated factory as a black box. Within the factory there is noise

pollution which has its impact on health. Disasters like the one at Union Carbide at Bhopal took place because of this approach. Electrostatic precipitators have been developed to contain the dust. Water discharge must be treated before it is allowed to go out of the factory premises.

5. *Packaging:* It can reduce damage to environment. Green movement tells the consumers to use re-usable packages. Recycling can be promoted by starting collection centres for plastic bags, used papers, etc. It is disappointing that trade unions don't take interest in environmental issues.
6. *Outward transportation:* How to minimize costs and transport by the least polluting means? Motor vehicles are carrying pollution deep into villages. Bulk transportation and use of local manpower may increase capital cost, but the process becomes less polluting.
7. *Product mix:* Environment friently production processes, and goods should be increasingly used. Tobacco companies can move to new products using the same skills. There is no point in moving a polluting factory in Delhi to some other place. It does not do any good.

Every company with Rs 100 crore turnover should have a senior Environmental Ethics Manager. Companies should carry out environmental ethics audit. To see how far they have moved in the direction of clean environment, Ethics Missions should be established. Interfirm comparison should be made to identify companies which have done well. Local, national and international awards should be instituted. Audit would show weaknesses. How to advance environmental goals should be included in the MOU. Environmental strategy has to be strongly built in the management system. Legal minimum standarts must be achieved.

From a complete denial of any responsibility the the 1960s in the USA, it came to the second stage of delaying and diluting even the minimum or notional compliance. In the third stage, it became a part of enlightened self-interest,. And now some companies have become pace-setters in forestry and similar other fields. And finally, many of them went further and got involved in environmental education and knowledge development.

I am of the view that it is in the interest of business to educate all concerned on how their life-cycle, etc., can help develop environment sensitive technologies and products. Sharing, monthly or weakly, with employees the environmental issues, down to the machine operator who might be misusing lubricants, would go a long way in conserving resources and reducing pollution.

Thus the business has to work in partnership with the people and environmentalists. Partnership with academia, the government, and chambers of commerce will certainly help. I am advising companies to work with the NGO's (instead of going to cold war) on how to achieve business objectives without having any deleterious impact on the environment.

Federations of chambers of commerce and industry should come out with codes of conduct for their members. Global partnership should lead to practices which discourage companies talking an unfair advantage by using polluting technologies or trading in polluting products.

Now let me put before you a few suggestions which, in my view, should go in the UNCED document in one form or another.

1. Environmental ethics will be a cruel joke on the poor unless we have a clear strategy to meet the basic needs of the people. Transcending national and regional boundries, we should use the world resources in such a way that the basic needs are met on a priority basis.
2. Resources should be protected from unprincipled reckless use; they should be conserved and used in a sustainable way. The North should compensate the South for remedial investment because much of the environmental destruction has resulted from the environmentally insensitive industrialization and life-style in the North. Severe standards for the relatively better-off parts of the world should be set to bring down ecological degradation in the future.
3. R&D work on environmental and related matters should not be confined to private initiatives. There should be a global cooperative endeavour. Population control, for example, needs a massive educational programme. It is a world-shaking problem; not a problem of India and China alone. Global initiative is needed for its solution.

Let us be clear that we cannot avoid sharing the globe as equal partners. Let us consider ourselves as a joint family. If anyone commits a mistake, let him repent and pay for it. There are principles like equity, natural justice, paying capacity, etc. We are now bound by collectivity; and mutuality between North and South. In fact, South will be bigger market in the 21st century. Let us, therefore, put our house in order and continue to pressurize the North to pay for the damages they have caused to the environment and ecology.

4. Gandhiji laid down the principles of trusteeship for the business wherein one holds a post or owns business as a trustee. Unfortunately in India business by definition had to be exploitative. Today in the world, the concept of business as an intermediary among various interests has gained ground. Thus Gandhiji was ahead of others in this respect too.

   Similarly, Gandhiji was for relevant technology; he was not against technology. Leadership for him was not a charisma. It was a duty; opportunity to serve *(Shrestha Dharma)*. *Yatha Raja Tatha Praja* (people follow the king). Days of the Rajas are over but in every society there will be the elite. Let this elite be a responsible one, a dedicated rather than a *parasitic* one.

5. Those of us who are in managerial leadership have the responsibility to show by example. Our behaviour governs the behaviour of others. In industrial relations, Gandhiji advocated a model of harmonizing the interests of management and labour. The futility of a permanent hostility between management and labour is being realized all over the world. Again he was far ahead of others.

Fusing of Scientific Temper and Ethical Values

Rajiv K. Sinha

Previous speakers have already outlined the basic considerations which enter into the evaluation of science and technology. When we consider ethics, we really do not have very soft and hard technological options for it. It is based on values systems that we have inherited or have acquired through education. The philosophical foundations of the value system have to be understood.

Thus the whole question of science and technology and environmental ethics is really a very broad question and hence one tends to become very speculative. The fact of the matter is that we have done those things we ought not to have done. We have left undone those things which we ought to have done. We have to start thinking about ethics, along with science and technology. We have tried to sweep under the floor the ethical dimensions. Some of these issues are of fundamental nature. In the process of development and modernization, the human values are always brushed aside. This is unfortunate, but true. The ethical ideas are considered to be static. The views of Gandhiji are considered antiquated, although many of those who think so might not have ever read Gandhi. His ideas have not become static; but the expressions he used may not be relevant now. It is our task to create new expressions to carry forward his views.

Gandhiji laid a lot of emphasis on means and ends. The achievements of science and technology have been very outstanding and we have made very original contributions in both the fields. They have really expanded the scope for human action and people have come to rely heavily on both for the shaping of their future. We have seen how we have probed the outer space and the deep seas. It is all because of science and technology. We cannot ignore the superior knowledge of cybernetics and computers. But we are not sure about the future. For that we have to look to ethics.

Global Issues in Science and Technology Policy

B R Nanda

So many interesting ideas have been thrown up. So many good examples have been given. It will take some time to integrate them into a framework. The dilemma is that science and technology are important; we cannot jettison them. In fact they control us and if we try to control them, we can succeed only in a limited way, if at all. Because, as I have said repeatedly, we now have a global economy which means that what happens in one country affects other countries.

We are not free agents. In fact human beings are not free. The day we are born, we become unfree. No individual or nation is, however, completely free. We know that the environment has to be clean and pollution free but the question is, how do we do it. That would dpend upon the situation in each country with specific problems and resources. I felt encouraged to think that some of the positive things which have happened in Europe, particularly in the EEC, would spread to the USA also, and thus the global situation, which looks depressing at the moment may change for the better.

It is not that the USA does not understand the problems of environment. They do. But they are under such great pressure from leaders in industry, business and politics that they are unable to match perception with performance.

Nehru was mentioned in the discussion as the pioneer in the science and technology in solving India's socio-economic problems.

As a Prime Minister he promoted scientific education and research. But, in my view, Nehru was far above his contemporaries in vision and thinking and after him his ideas were neither developed nor properly followed up.

Nehru wanted the universities and national laboratories to be integrated with society. He was against their isolation. And if one reads the letters he wrote to the Chief Ministers, he, time and again, emphasized the openness of these institutions and regretted that national laboratories were getting bureaucratized and universities politicized. Both were getting isolated from industry, agriculture and the common people.

I came across an extract of a conversation between Norman Cousins, editor of *Saturday Review* (USA), and Jawaharlal Nehru. Most of the American newspapers were unfriendly to India during the freedom struggle. But *Saturday Review* was an exception. Cousins asked Nehru "What do you think about the debate on determinism and human freedom, i.e. faith and human freedom"? Nehru: "It is a complex question you are asking me. I think we can reconcile the two". Nehru asked "Do you play bridge"? "I do play sometimes. But I am not a very good player", said Cousins. Determinism is the card that is dealt to you; human freedom is the card you play. No body can be totally free. The card has been dealt. Science and technology dealt the card we have. Now how do we play it? It is part of the human freedom.

It is interesting that science not only creates problems but also helps us to solve them. It is science which has given us knowledge about the ozone layer depletion. It is now up to us, to our ethical and moral values to use the scientific and technological card for the benefit of mankind.

It is in this perspective that I consider this Symposium of seminal importance.

Both Science and Religion Need Transformation

R P Misra

As we proceed through the Symposium, we become clearer about the issues involved. Yesterday we thought that marry of our values were derived from religions Hinduism, Buddhism, Islam, Christanity, etc. We put the blame on science and technology. Now we realize that even though science and technology have caused a lot of damage to the social fabric; religion also has not been very much behind even though in different ways. If science has to be retrieved from its unethical framework, religion too has to be retrieved from its unscientific temper.

We are passing through a period of confusion in all the aspects of human life. Why blame only science and technology. Every social discipline wants to be suffixed by science. So we substitute the term humanities and social studies by human sciences and social sciences. We now have library science. Then tomorrow the term religion may be replaced by religious science.

At least in this country, we firmly believe that whenever humanity faces the challenge of survival, a way is found. This symposium is an endeavour to move in the direction of finding this way.

Towards a New Paradigm

G L Pandit

I think we are in search of a new paradigm. The search now for environmental ethics does not mean that there was no environmental ethics earlier. In my considered view, all ethical values impinge on the environment and hence environmental ethics is only an extension of the ethical values of society.

Our knowledge about the environment has vastly increased. It is impossible now to ignore this knowledge. There is also enrichment of the ecology of human conciousness, our awareness of ourselves, of others, and the whole ecological system. Science and technology now play a very important role. This has been a mixed blessing of progress and regress.

We must try to understand the logic of the situation which calls for an ethically sound human behaviour. The situation tells us that while there has been progress in science and technology, serious mistakes have also been committed even though unintended. The ozone depletion is one example; overpopulation is another. These mistakes have accompanied the progress in science and technology. Progress and retradation have gone together. They are inevitable and indivisible. These mistakes have been committed everywhere. We cannot say that they were committed there and not here. We should not blame the West alone.

The greatest task before us, therefore, is to identify our own share of mistakes. If a disaster takes place somewhere in the world, we are partly responsible for it. We cannot absolve ourselves by saying that we were not there.

On our side, in India, what we should do is to identify the mistakes that we have committed. I remember Prof Gangrade talking of limits of economic growth. This reminds me of the problem of population. There is a view that came forth even yesterday that overpopulation is not so bad. I think, it is a mistaken notion. There has to be a limit to population otherwise we will one day be like sheep fossilized in a forest with grasses. What ethical values would help us bring down birth rates, should be of great interest to us.

Let us look at our life-style. Our emphasis on natural living, *Yoga*, Ayurveda, meditation, and so on call for an appropriate attitude towards the Himalayas because we need serene, calm and pure nature for meditation, we need medicinal plants and clean water. Some of us may consider them outmoded, unmodern, etc. but look at the speed with which they are being adopted in the affluent societies of Europe and America. This value is linked to the natural beauty and natural environment.

Let me now talk about an aspect which provides the scientific foundation for environmental ethics. We call it the law of ecological harmony according to which information lost in one embodiment is information gained elsewhere. It has also implications for interaction among plants, animals and human beings; among place; and among peoples.

If we do now identify our environmental mistakes then we will not be able to prepare ourselves for the future. If we continue to be careless, we may face a big catastrophe for violating the law of ecological harmony. If we act in such a way that information lost is not gained elsewhere then we will be like fossilized sheep.

In defining the goals of science and technology, we should join hands with scientists and technologists. We have to be truly interdisciplinary for the environmental problems we face today are not disciplinary in nature, content and implications.

M K Machwe

The deteriorating environment around us is posing a global challenge. A number of factors have contributed to create the feeling of fear and concern that the Earth is an endangered planet. Some of these factors are:

uncontrolled growth of human population, widening gap between rich and poor, deteriorating situation of resource availability, environmental degradation leading to pollution of land, air and water, nuclear radiation, Green House effect, ozone layer depletion, etc. All these factors are interrelated and could be directly lined to the progress in science and technology. Science and technology have been blamed for all these evils. But I would like to say that science and technology have also produced the green revolution without which millions would have died of hunger.

I would not like to enter into a debate on the usefulness of science and technology. I would only like to endorse what Harvey Brooks said in this connection. He said: "There are those like myself who think that technology is absolutely essential for survival; and there are others who think it is the opposite. It seems to me that because technology opens options, it takes a strong value system to control the choice of options." Science and technology have opened new options which were not available earlier. Everybody was moving on one track. And it was science which has given the options. Hence if one wants to control technology one needs a strong value system. The technology is disasterous when it is adopted by a society with a weak value system and very little sense of collective welfare. Science and technology is value neutral and cannot be called upon to decide between preferable and pleasurable options available to them. Science can tell us what can be done and what cannot be done but it certainly cannot tell us what ought to be done. That decision falls within the realm of ethics. It deals with the moral principles, moral duty of man, and with the question of right and wrong, standards of conduct, etc.

9

EDUCATION AND ENVIRONMENTAL ETHICS

9

EDUCATION AND ENVIRONMENTAL ETHICS

Opening Address

R P Misra

Education in its broader sense covers all that man learns through various channels of communication. It is innate for a plant to grow in a particular form; an insect to behave in a particular way; and an animal to be no different from other animals of the species. Flower blooms, lions roar, elephants are vegetarian whether they are wild or domesticated. Man, the supreme creation of God, or Nature, or chance survival of the fittest, will not be man in the form we see him today except in a society of humans. He alone among the living things needs prolonged rearing and education to be human. This is the case not only with the modern man but with the most primitive Jarwa of the Nicobar Islands too. Jarwa is Jarwa because of his upbringing in Jarwa family. There have been cases of human babies being brought up by wild animals. Such babies behaved like their animal parents. Perhaps the first primitive man began his life like that. It must have taken several millenia to chance acquire and accumulate the experience to walk on two legs, to communicate by a sort of language, and build something like a cave or a hut. Much of what man is today is because of education rooted in rich tradition, history, knowledge, culture, etc., etc. What is innate is very very limited.

Modern man has such a long written or unwritten history, such a far-reaching knowledge base, and such a rich cultural experience that by the time he is adult, he is already entrenched in a particular language, religion, culture, society and environment. He looks different from others not only because of heredity; but also and perhaps more importantly because of different educational processes he goes through.

What man acquires in the early age of his life shapes his character. What he acquires is through the process of integrated education. He is not taught disciplines like physics, geography, management, philosophy, and engineering to mention a few. He gets a wholistic knowledge. Various aspects of life and environment are related to each other. He experiences and learns in bits but in relation to the whole. Parents and neighbours are his first teachers.

Then at the age of five he goes to school. *Locomotion* is fascinating for a child of that age. Every child wants to be a cartman; a railway engine driver; a pilot of aeroplane, depending on the context of his birth. Urge to see more, to command and to come of age is persistent. His contacts widen to give wider experience. Teachers teach him different subjects. Knowledge starts getting divided broadly, to begin with, and then in a variety of subjects. Other boys and girls help him learn what he could not get from his parents, good or bad. Teachers have a great influence on children. They look to be all knowledgeable; people to be emulated.

Soon the child either gives up for on reason or the other or he becomes a specialist. If he does not go to school, his knowledge is wholistic but too limited; if he goes to school, he learns more of less things. He feels lost; but he graduates to become a specialist. He is no longer just an educated man, he becomes a mathematician, a historian, a physicist, or the like.

This indeed is the dilemma of education. It does not make man human, it makes him specialist. So those who have no formal education become unfit in this complex world; and those who have formal education tend to create a world which is self-centred and fractured.

One of the adverse impacts of fragmented education is the ecological imbalance and environmental pollution. It has its roots in education. Its solution also, therefore, lies in education, in its content as well processes. Man's attitude towards nature and environment will depend on the type of education he receives. This has necessitated the call for environmental education which is being promoted at the global level through UNESCO-UNEP International Environmental Education Programme (IEEP) since 1975 and other agencies and at the national and local levels through various instructional and mass media channels.

Environmental education has two objectives: to promote environment sensitive behaviour, i.e. environmental ethics, and to provide knowledge about the causes and consequences of environmental pollution and degradation. It has to begin from early childhood. It should really become part of the aculturization process. People in the pre-industrial societies were sensitive to nature and environment because they were so educated from early childhood. This education was through words of mouth as well as through live examples.

To create a similar situation, we have to introduce the study of ecology and environment at all levels of education and for all sectors of population ranging from children to top decision-makers. The most important target group of environmental education should be children, the citizens of tomorrow. They should be brought up in an educational environment which fosters a healthy man-nature relationship. Adults should be sensitized through mass media and other relevant channels like conferences and Earth Summits for they constitute the parents who are the first teachers of children and who lay the foundation of environmental education in the minds of children. All our educational resources have to be involved in this exercise to redirect, if not reverse, this trend of almost criminal attitude towards nature and human habitat.

Participatory environmental education is an effective means to sensitize the people and change their environmental behaviour. Naturalist groups, aesthetic and fine arts clubs and associations, gardening, walking, cycling, etc., are among the important ways to cultivate environment friendly behaviour, and to create sensitivity towards nature and ecology. These and other approaches help people find an answer to the question: How to deal with others living with mankind on this beautiful Earth.

As regards higher education, there is need to reconsider basic premises and attitudes in different fields of studies. Creation of scientific and ecological knowledge is essentially a task of university education. It gets into action at the people's level. At the high school level, biology and social studies with the focus on geography, history and philosophy can effectively integrate environmental knowledge and ethics. At the elementary level, children must be sensitized and trained to deal with their environment in a positive way by developing interest and respect for other forms of life.

Environmental Ethics Is the Goal of Environmental Education

A Ghafoor Ghaznawi

In UNESCO we have an Environmental Education Section which has implemented since 1975 the UNESCO-UNEP International Environmental Education Programme (IEEP).

Environmental education has its objectives to promoted awareness, to provide knowledge and to inculcate attitudes, values, commitments, and ethical responsibilities in its various target groups at various levels, for the rational utilization of natural resources, and for the protection and improvement of environment. Thus you can see that ethical responsibilities or development of environmental ethics is one of the objectives of environmental education.

I would like to dwell upon a few words we often use. It is necessary because in a discussion like this, the words/terms we use should mean the same thing to all participants. There should at least be an agreement on the meaning. Let us take the word "environment" first. In our thinking environment is the sum total of all external factors that affect us. In its holistic frame of reference, environment thus includes natural as well as man-made phenomena-physical, biological, economic, political, cultural and other dimensions. Thus our concept of environment is holistic and interdisciplinary.

What should we then include in environmental ethics? I have been thinking about this for quite some time. Now I am glad that an acceptable definition emerged. I have defined environmental ethics as an internalized code of behaviour for guiding and controlling human actions in the presence and or absence of environmental laws for understanding, protection and improvement of the environment. That means that within the man there is a force that needs to be sharpened. And once your sharpen that force, here in the context of environmental protection and improvement, it controls human behaviour and action. It should be considered in individual and community or social context. This is how I feel about this.

The need of environmental ethics is very clear. It fosters the development of environmental laws, attitudes and behaviours. It is to supplement existing environmental laws and legislation, it is to control from within human behaviour and actions on various aspects of the environment in the presence or absence of laws. Thus we are dealing with a very important aspect of controlling environmental planning and management.

Environmental ethics plays an important role in the planning and execution of local, national and international projects. Construction of dams, cities, airports, highways, etc., was ethical implications. It has a lot to do with business, even export and import of materials. For example, when you export to some other country the materials you do not want to use yourself, there is an ethical issue. If it is not good for you to keep your toxicity at home, is it good for your neighbour? You pay them so that they buy the danger, knowingly or unknowingly. It is an ethical issues. CFC emissions are now threatening the ozone layer. The environmentally friendly substitutes being made and given or not given to others on political or economic grounds involve ethics. Certain medicines are being tested on the poor. Isn't it an ethical issue?

What are the roots of environmental ethics? We in the UNESCO have found out that environmental ethics is present in most cultures and religions for the purpose of harmonizing humanity-environment relationship; or to bring humanity to terms with environment. A number of studies have been summarized in an issue of *Connect*. Prof Misra quoted it in his opening address. UNESCO did a Consultation Meeting on Environmental Ethics in Cairo in 1992. The report of this meeting is available with the Gandhi Bhavan, Delhi University.

The question is: how to inculcate environmental ethics in the people? Various mechanisms have to be considered. In my view, it has to begin form home. Here it is the parents who should bear responsibility of giving the environmental values and ethics to the children. Among the parents, the role of the mother is fundamental. She is the best teacher; child is the best student; and home is the best school for learning. Her teaching methodology is interdisciplinery. The mother does not divide her instruction into chemistry, physics, biology, etc. It is the whole life and it is the whole environment. And to the child, environment is really of wholistic nature. It is one unit, not divided in separate units. It is, therefore, easy to teach the children the wholisticity of environment and why its protection is a must. Mothers must, therefore, be educated for this purpose.

The second mechanism covers various types of religious institutions. Human beings are inducted into one religion or the other the day they are born. Fortunately, almost all religions incorporate environmental ethics in their codes of behaviour. Therefore, religious institutions are important in imparting environmental values and adapting them to today's needs. We have to develop programmes with these institutions. I can mention our experience in this regard in Egypt. The Ain Shams University of Egypt was very courageous in sending to us a proposal to train the *mullas* of the mosques of a region in environmental ethics. It was in 1990. I was pleased with the proposal and the courage. We supported it. Fifty *mullas* of mosques from Cairo and its surroundings were given lectures on how the environment and Islamic teachings were interwoven. The importance of environment in Islam was elaborated by Prof Raza yesterday. It was possible to convince those people to participate in such a programme. The question was what should the *Mullas*, after training, tell the people and how. They were given as assignment after they learnt certain concepts of environmental protection, sanitation, etc, to prepare a message, which they could announce on Fridays before the prayer, the mullas *speak* in the mosque. This was implemented. The feedback information we have is that the exercise has been quite fruitful. Such attempts can be made elsewhere involving priests/preachers of various religions.

The third mechanism is the mass media. The mass media has to prepare and communicate meaningful and accurate messages for the general public. Among them include the concept of environment, understanding that man is an integral part of environment, that he depends on the environment and hence he should come to terms with it, that man should improve and protect the environment. The media should revive the cultural values relevant to environmental protection and management and focus on its inculcation through appropriate seminars, round tables, reports and even rewards.

Lastly, I would like to mention the role of formal and non-formal educational systems in the promotion and inculcation of environmental ethics. Here the current curricula have to make room for environmental ethics. Teacher education has to make a room for environmental ethics. Then the teacher has to become a model. Unless he/she becomes a model for the student,it is very difficult of for the students to change their behavior. The teacher has to live up to what he says. Evaluation techniques and instruments are to be developed to know if our efforts have achieved the goals of environmental education, i.e. inculcating environmental ethics are the students. As its development takes time, its evaluation would also take time. Besides it won't be so easy. Therefore, we have to develop a propriate impact assessment measurements.

Mrs Zena Daysh

Thank you Dr Ghaznawi. In my thinking it could not be better said than the way you put it.

The four channels that you mentioned cover the whole gamut of environmental education to inculcate environmental ethics. The first channel is the mother, as the foremost teacher; the second is the religious institutions; the third is the mass media: the newspapers, magazines, radio, TV, music, dance, etc., and finally the formal and non-formal educational institutions ranging from mursery schools to universities.

I will go a step further and add a fifth channel, i.e. the leaders in politics, in decision-making, and in industry and in science and technology. These are the kinds of persons who would assemble in Rio for the Earth Summit on Environment and Development in June. The Presidents and Prime Ministers of so many countries are going to put their heads together. What are they going to do there? Scientists in London are already saying that nothing is going to come out of it. This is going to be *White Brãzil*. May be it will be *White Brazil*. But what I am more concerned about is what goes up to Brazil and what comes out of it. If we are ready with a clear message before UNCED, I am sure something meaningful would emerged from it. It has back and forth insight. I urge Dr Ghaznawi to look into the politics of environment—politics not in a small sense, but in a sense of the art of compromise and movement forward. Political judgment, and let me say the ethics of politics, must be constructively used to prepare the message.

A Ghafoor Ghaznawi

Thank you Zena for the fundamental question you raised. I did not touch this fifth level, i.e. political level. But in a way I did this when I said that the need for environmental ethics existed in the preparation and implementation of projects like dams, highways, etc. Here it is the decision-makers who are involved. They have ethical responsibilities while making decisions on various development projects and issues. I referred to the export and import question. Here, commerce, business, political leaders are involved. They have to exercise ethical responsibilities. I also said that if DDT is bad at one place, it cannot be good at another place. It has ethical implications and responsibilities. If we dump toxic waste in our neighbourhood we are unethical. In industry I referred to CFC. Who is responsible for it? If someone has got replacement for CFC, is it ethical to price it so high that most countries cannot buy it? How to handle the transfer of environmental friendly technology? There is an ethical issue. Unless we are imbued with the feeling of sharing at least in the field of environmental technology, we cannot make progress.

But how do we do this? The leaders of tomorrow are in the schools and colleges. They are in the laps of mothers. If we provide a right kind of education, we can be sure about what the future leaders will be and do. For the present leaders, mass media, the conferences like UNCED in Rio and bilateral and multilateral negotiations, United Nations, and the pressure of the NGO alone can help.

Thus, I fully agree with Zena that the politicians and decision-makers need to be educated particularly in areas such as environment. Their knowledge about environment is either limited or superficial. These people are very busy, and, therefore, they have to have advisors around them for briefing and for providing information and advice for fostering sound decision-making.

Intrinsic and Extrinsic Values and Environmental Ethics

T S Rukmini

Environmental education needs revolutionary changes at the political level. But we know, politics has never been revolutionary. It is the people who create revolutionary changes. And hence it is at the grassroot level that much of our effort in environmental education has to go.

A Ghafoor Ghaznawi

I said yesterday that an informed and well-educated public will take a right decision. And our responsibility is to inform the public; give them as accurate and appropriate knowledge as possible. In a democratic society, they are going to vote for their leaders and decision- makers.

N Hösle

I want to make some remarks on ethics, education, and environment. I will begin by making two distinctions. The first distinction I will make is between intrinsic and extrinsic values. The actions which we regard as good in themselves are actions which are by themselves important but whole consequences can have great ethical implications.

Let me give an example of an ethical decision. "Not to lie" does not have an intrinsic value, as honesty is the basis of normal society. But can there be situations in which the decision "not to lie" has consequences which we evaluate negatively. I think the major problem of ethics is always to come to a middle way between ethics with a strictly intrinsic value and consequential approach which recognizes that many attitudes which are well intentioned and are based on purity can be adopted. This in my view applies also to environmental ethics.

I want to introduce another distinction. There are fundamentally two different arguments. One argument says that future generations have a right. If we pollute environment we are depriving them of their rights. It is really a matter of metaphysical interests as to whether to recognize the rights of those who just do not exist. We have duties to preserve the nature for future generations. Nature is the basis of survival of mankind and so we have to preserve nature. Another approach says that it is not a question of survival of mankind in this or future generations. Nature in itself has to be respected. Even if we cannot prove its utility for mankind, it has to be respected and protected.

Since we are here both from Asia and Europe, I say that our attitude to nature depends on our cultural background. It is interesting to remark that your tradition, the Asian tradition, always had a higher respect for nature than the European tradition. There is little doubt that in modern scientific enterprize, nature is lost and, therefore, I have great sympathy with all attempts to go back to cultures and religions which have a deeper feeling and empathy for nature.

On the other side, I am convinced that all the ecological problems we face today cannot be solved by science and technology, and this seems to me the main reason for emphasis on environmental ethics. While we want to perfect our technologies to cleanse the environment, we also want to have a change in our value system with regard to nature. Both approaches must be complemented. It is possible to have a well functioning ecosystem technology which would overcome the worst consequences of science and technology for nature but such an approach will deprive human beings of the pleasure of having a normal loving relationship with nature and fellow human beings.

I will insist that only a revival of the traditional loving behaviour with nature is not a solution to our problem. For example, to be kind to a cow is indeed of very great value (which the Western tradition has now lost). There are some deep insights into the area of rebirth which seem to express the unity of man with other living organisms. But all these ideas will fail in solving our problem unless they are accompanied by scientific and technological approaches. I am not sure that such an enthusiastic appraisal of traditional relations with nature will address the whole problem of environment. The *mullas* of Egypt being trained to address the environmental problems is interesting but of little consequences in my view.

The pressure of mankind on the organic nature depends on number and we will not achieve much if we do not address the demographic behaviour. Traditional religions have value systems which cannot internalize population control. And only insistence on the beauty of the traditional way of living with nature without a clear analysis of the necessary consequences of our demographic behaviour, in a world in which a child has dropped? In which our life expectancy has become much longer, will not solve our problems. Therefore, it is very important to bring the traditional and scientific approaches together, in order to solve the problems. We should try to avoid, on the one side, the traditional romanticism and, on the other, the cold technological approach.

With regard to the educational problem, I will insist that environmental ethics is always a personal model, much beyond the theoretical discussion. Prof Poser yesterday distinguished between the foundation of a norm and the way this norm could be implemented. Here the concrete model of how we live will have much a stronger impact than any laws, discussions and publications. Again in this model, we must try to address the fundamental issue of the technological age. This age has a new situation of ethics because the consequences of actions are much extended in space and time. I do not want to discuss the nuclear issue but it is obvious that we reflect from that stand. If in future half of plutonium that we have today remains there we assume a responsibility which no former generation had ever before. The actions of future generations in space and time will be infinitely greater than ours. We did not succeed in developing a set of ethical norms with a corresponding increase in our power. The traditional ethics is based on lifesize approach; one could see immediately the results of what one did. Modern technology has empowered us to do things the consequences of which we cannot see. We cannot see nuclear reaction; we do not see the ozone layer depletion. We need the complex scientific measurements. All this needs a change in the ethical norms of mankind. I am convinced that if the development of scientific and technological rationality is not equally matched by a new development of ethical rationality, we would not solve our problems.

A Ghafoor Ghaznawi

Thank you very much Dr Hösle. I think the trend of the discussion has been in favour of your views. We must take advantage of the progress humanity has made. If this progress is proving harmful in some respects, we must evolve counter and balancing measures. Environmental ethics plays this balancing role vis-a-vis science and technology. The answer lies in dovetailing the two so that the progress we make is not only in the right direction but also sustainable.

Women and Environmental Education

R B Singh

There is a Chinese saying which goes like this: "If you are thinking one year ahead, plant the rice. If you are thinking ten years ahead, plant a tree; and if you are thinking a hundred years ahead, educate the child," This shows the role of education in environmental awareness building process. Awareness leads to action.

Overemphasis on formal laws for protecting the environment in my opinion is not very helpful. While hard laws do have their roles to play, it is often the soft laws which emanate from ethical values, and which can also be referred to as conventions which help the people become positive to environment in the day-to-day business of life.

This is particularly important for countries like India which have a democratic system of government. It is very difficult to force anything on the people. Appropriate and adequate public opinion has to be created for even a very good legislation to be accepted and complied with. It is here that the role of mass media is really crucial.

Women constitute half the population. In India their status in terms of social recognition, education and health is not good when compared to that of men. And this situation more or less prevails in other countries of the world also. While formal educational facilities need to be expanded and girls encouraged to attend school, it takes time before the benefits of education are appreciated by the illiterate parents. Here again the role of non-formal education and mass media becomes critical.

We should also look at the issue from the human rights angle. Human rights constitute part of ethics, for it is the recognition of certain human values which give rise to human rights. If science and technology work against the fundamental rights of the individuals particularly the poor, necessary amendments are a must. Thus the human rights should become a part and parcel of environmental education, formal as well as non-formal.

Let me also mention here the need for an international convention on Environmental Rights and Duties. Such a convention would go a long way in educating the people about the need to have a more positive attitude towards nature.

Professional Ethics and its Environmental Implications

K D Gangrade

We can put our three days' discussion in terms of three Cs: conservation of knowledge, creation of knowledge and communication of knowledge.

By conservation of knowledge I mean to say that our past experiences contained in our cultures and scriptures should be examined in the light of the present scientific knowledge, and revived and utilized to solve our problems. New knowledge must be created by deeper studies, field work and experiences. And it must be communicated through various educational channels.

Education has two major components; one for the individual and the other for the society. I am very fond of *Panch Sheel*, the five points of our education? *First*, cultural transmission. When I use the word conservation, I mean cultural transmission to coming generations. *Second*, social cohesion and harmony. I think, we should aim at perfect harmony between nature and man. *Third*, innovation it is a must for progress. We cannot just go on importing technologies. We must build on whatever we have through innovations. *Fourth*, selection and allocation. *Fifth*, personal development.

Most of us assembled here are educationists in one way or the other. I got inspiration from Prof Chaturvedi to say that there are three kinds of teachers. One tells and lectures, he is average; second, demonstrates; and third, inspires. But a teacher should not only tell, demonstrate and inspire, he should also put what he says in action. Education cannot and should not be confined to the four walls of educational institutions. It should be imparted through (1) family, (2) school, and (3) community (local and global). If all the three converge on the common goals of education, then the ethical dimensions can be well taken care of.

Let me say with a sense of honesty that the so-called intellectuals are the greatest defaulters in matters of ethics. I am told that the city of Chicago consumes more energy than India as a whole. But it is our intellectual friends from the USA who teach us the need to conserve energy, etc. Humanity is one in more than one sense and hence each one of us has the right to tell others about the values of environmental reconstruction. But we have no right to tell others on what to do, unless we ourselves practise what we preach.

What Can UNESCO Do?

T K N Unnithan

Environmental education must incorporate environmental ethics. That is the only way we can make individuals conscious of their moral responsibilities. True, we should point our finger to anyone who pollutes the environment but we should also look at ourselves as individuals and as groups.

Let us not restrict the definition of environment to nature. It should be seen in broader perspectives to include human beings. Human relations have a lot to do with the quality of the environment. Wars destroy all that comes in the way; they consume so much of natural resources within a short period of time. So, our responsibilities towards other human beings constitute important dimensions of environmental ethics.

Several years back we started the programme of value education. Several Vice-Chancellors met and made recommendations. That is now contained as Point 8 in the 1986 Educational Policy Statement. But since then nothing has been done except for some more seminars.

How to carry the message of environmental ethics to the children at various levels of education, kindergarten to university. We must re-examine curricula in this country to find out the place environment gets in them. At present it is totally absent. Would UNESCO organize a workshop for this purpose? And then follow it up to see that it gets implemented. I do not know what the NCERT can do in this regard.

A Ghafoor Ghaznawi

India is one of the countries in the forefront to have an official mandate through the National Education Policy-86 to incorporate environmental education in the full spectrum of education. The challenge to translate this mandate into action and to carry it to the grassroot level is a long-term undertaking with challenges for all educational institutions.

UNESCO has worked with the National Council of Educational Research and Training (NCERT) and the National Institute for Educational Planning and Administration (NIEPA) in India. We have worked with Delhi University, and we have worked with several other institutions in the past towards this goal. We are delighted to be here with CHEC and Sustainable Development Foundation today.

I share the view of Prof Unnithan that we have to carry out research on how the recommendations and suggestions made at various time and levels for incorporating environmental ethics in the curricula are being followed up and taught appropriately. UNESCO will be delighted to study for consideration such proposals, if they are channelled through the National Commission for UNESCO.

10

DEVELOPMENT STYLES AND ENVIRONMENTAL ETHICS'

10

DEVELOPMENT STYLES AND ENVIRONMENTAL ETHICS

Opening Address

R P Misra

Development style and quality of environment are closely interrelated. Attitude towards environment is built in the very concept of development. Different styles of development produce different patterns of man-nature relationship.

Development and economic growth are often treated as coterminous. While in theory they are distinguishable, in practice they still mean the same thing. Development aims at raising the income of a nation and its inhabitants. Of course, it aims at education, health and other welfare measures too, but only through more production of goods and higher income.

Higher and larger scale production needs a supporting market, Promotion of production and consumption, therefore, goes together. Modern science and technology helps, in mass production of goods and consumerist culture helps in the consumption of the goods so produced. This is the growth oriented development style that now prevails all over the world.

This style of development pervades all walks of life. There is stress on quantity, not on quality; on income in money terms, not on quality of life; on standard of living, not standard of life, on money value, not on use value, and so on. Success of a nation or individual is measured by income irrespective of how that income is generated and how long would it be possible to maintain that income.

Much of the environmental problems we face today can be traced to the style of development being pursued in the world. Intensive use of modern science and technology to exploit the nature for more goods, without being concerned about its ecological consequences, has now landed us in serious environmental troubles. While our goal was to reach newer heights of happiness, we now find our future uncertain.

Man's craze for more goods and more power to control and exploit nature led him to control and exploit other human beings too. Renaissance and Enlightenment in Europe aimed at liberation of man from the so-called bondage of religion and tradition. It declared the equality of man, but this equality remained confined to Europeans and their kith and kin. The black, brown and yellow races were not considered equal. In fact, the "savages" were eliminated in thousands and other non Europeans were subjugated to yield natural resources, raw materials, slaves and *indentured* labour. They did not have the same rights as the Europeans. So while the Enlightenment stood for freedom and equality of man in Europe, it ushered in an era of darkness for the rest of the world.

But the spark that Enlightenment even in its limited scope caused did ignite new hopes among the elite the world over. One of the great achievements of Enlightenment was to put man as the measure of everything. Man was put in the centre giving rise to an ethnocentric view of the world where in the non-human world was nothing but are source for the use of mankind. It was a great departure from the physiocentric worldview. While the former authorizes man to appropriate the world for its own self-aggrandizement, the latter argues that mankind as well as other conatural beings must co-exist. According to this thesis, we humans can use nature only to the extent that we contribute to it. We have no right to jeopardize the existence of other forms of life beyond nature's capacity to recoup.

The Gandhian Alternative

T K N Unnithan

The world is changing fast. Western Europe is taking the form of a nation-state; Soviet Union has distintegrated; Communism has failed as an ideology.

What is happening to America? It is suffering from an absence of enemy. The enemy is gone. What to do with the war industry. American economy is dependent on war industry. We are, in this country, being dictated by the powers that be to replace one wrong by another abandoning all that we stood for since Independence. The Gandhian model which offered an alternative has been forgotten.

Nehru borrowed his socialist model from the USSR. We talk like parrots about planning. Now Nehru's model is also disputed and discarded. For this country and for the world as a whole there is now no other alternative but a Gandhian model; not the one that traditional Gandhians love, but the one which is dynamic and always new by experimentation.

Gandhians are unwilling to experiment. For them nothing has happened to the Gandhian thought after the Mahatma died in 1948. Unless we take Gandhi out of the clutches of these groups, there is no salvation.

We have to be self-reliant, and self-sufficient. As Shri Kamal Nath said we should not follow the developed world in everything. We should chart our own course. But we cannot do this unless we fall back on a wholistic yet experimental model. We should have the courage to say that neither capitalism nor socialism will meet the challenges of the future. Only the *Sarvodaya* model of Gandhi appears close to this formulation. In my view, it is the only long-run alternative not only for the developing countries but also for the world as a whole including the USA and Europe.

Environmental ethics has to be looked at from a larger perspective. Environment is not just the physical environment. It includes human beings and the cultural environment. How people interact with each other is part of the environmental cancern.

K R Dikshit

Early in the history of man, natural environment was all-pervasive. Natural resources were abundant and man was lost in wilderness. His technology was primitive.

In those days, lots of people were born but only a few survived. That is why, India developed a fatalistic philosophy of *Moksha*. I do not believe that Indian society was ever in harmony with nature and within itself. And whatever social equity we had is easily revealed by the *Manu Smriti*. So not everything was rosy in those days.

The ethical values also change with time. We do not, therefore, have to dig out values from the past. We have to evolve new values as new demands and needs arise. Slavery was justified in Greek history. We also know that while the colonial countries had democracy at home, they practised colonialism abroad. Indentured labour was taken to distant islands to cultivate sugarcane. This is how the African population migrated to the USA. We also know how the kings were autocratic, and took the place of dieties. So let us not all the time think of rediscovering old values. I would not like to get into controversy in this regard but I must say that what various religions have said on many issues and what has gone in the name of religions is not something that we would always like to cherish.

Technological developments ultimately made man so powerful that he started overharvesting nature. The old equilibrium was disturbed and man lost sight of his being a part of nature. The dangers inherent in such an approach to life became visible and hence now the concern about environment and ecology.

We have now come to a stage where human species benefit at the cost of other species. I do not believe hat ethical principles are eternal and inherent. If they are so, they have to be modified in the light of new challenges merging from our material prosperity. I agree with the view that social consciousness and ethical values have not grown with the new challenges of our society. It may now be the time to do so.

As regards development styles, I see two major categories in the West. The first pertains to the capitalist world which is marked by consumerism, opulence and free enterprise, and the second to the socialist world which no longer exists. This style is marked by central control, command economy and State as the prime economic mover. Both rely on science and technology. In my view even in the former, there is a lot of social control. The modes of development were not much different excepting in the matter of ownership of the means of production.

The third option is being talked about in India. We had a great soul like Mahatma Gandhi who handed down great eternal truths and values. But to my mind, to develop India on the pattern of Gandhian economics is certainly no compatible with the modern world. I have heard people saying that we should have a simple style of life. Villages must be independent. I really do not know if Gandhi could have imagined this burgeoning population; nor could he have imagined this technological development. The question is, would we be able to take care of this population by measures of austerity. It is just not possible and feasible. Gandhi identified himself with the poor and he lived like a poor man. We do not practise dictums. The problem is that in the present set-up even in a simple lifestyle situation, agricultural production has to go up. Can an efficient and highly productive agriculture be supported by austerity? We need fertilizers; we need irrigation; we need hybrid seeds. We certainly need all kinds of new inputs. Such inputs would not come from a simplistic life-style. We need for their production scientific laboratories, factories, plant breedings, etc. And these all require a very complex type of social institutions and scientific organizations.

Every poor man uses antibiotics. He may not always need them. He may even misuse them, but at certain critical moments he cannot ignore them. What was the death rate before the world discovered antibiotics? The growth of population speeded up when antibiotics came. If we want to take care of the people in India even at the minimum level of living, education and nutrition, we cannot do this by the retrogressive steps being suggested here.

I think we have no way out. We have got to depend on more and more of science and technology. There is no point in complaining about what others are consuming. That is a fact. No body asked you to be poor and not to consume more. It is your choice. Why then grudge others being rich? Options are few. We can make some modifications but we cannot do away with the achievements of science and technology.

If I have to place modern science and technology, machine and factories on the one side and a simplistic living, and village independence on the other, I would choose the former. If any country wants to be independent at the village level, one has to be poor. I think the time for such fancy ideas has passed; we have missed the bus.

S Jenamani

Modern civilization and annihiliation of nature have gone side by side. There are two views about man-nature relationship. This is obvious in this Symposium also.

I have been doing research on perhaps the poorest and environmentally most degraded district of India, i.e. Kalahandi in Orissa. I would like to emphasize here that the poverty and deprivation of the people of this district is because of environmental degradation caused by the modern approach to development.

I learnt here that shifting cultivation which has been branded as a villain was not really so bad. It is based on sound ecological principles. As such there is a lot of scope for learning from traditional societies. The values they cherish need further investigation by all of us.

D N Chaturvedi

I would like to respond to the remarks of Prof Dikshit. I am a student of Gandhian economics. I have not seen so far any Western development model. What Prof Dikshit calls development model is really the economic growth model. The two are not synonymous. Science and technology are the instruments of growth. What I would like to say is that it is this growth model whether free market or socialist that has generated the present distortions.

The growth models are not concerned with morality, ethics etc. The moment you juxtapose these models against ecology, ethics and human values, the latter change and such a change produces such development models.

Where is the growth model which will give food for all, shelter and clothing for all, education and health for all and a creative environment for the development of innate faculties of all men and women to the fullest extent. The Communist model is already gone; and the Capitalist model has come to its last leg of journey. And now for an all-round human development the *Sarvodaya* model of Gandhi alone is the answer.

Natural Resources Constitute the Real Wealth

R B Singh

We are facing a great crisis. World is becoming ecologically unstable, socially alienated and economically nonviable. These problems have two dimensions: ecological and socio-economic.

When we examine these issues in depth, we are not able to establish a cause-effect relationship without a sort of value judgment. Population, pollution, poverty, and a host of other problems are so intertwined that it is difficult to discern which is the cause and which is the effect. Let us take the example of forests. Deforestation is not a recent phenomenon. Nor can it be said to be always bad. If human beings have to live on this earth, forests will have to be cleared, as other resources would have to be used. But the question is how are these resources to be used so that the delicate balance in nature is not disturbed leave alone destroyed to the detriment of life on earth. Part of the problem of environment can be handled by environment friendly technologies, but the technological approach has its own limitations. This approach must be backed by an ethical code of conduct which is spontaneously adhered to be the people.

The roots of environmental problems lie in the style of development, otherwise known as model of development. The growing demand for resources to meet the basic needs of food, shelter, education and health of the poor, on the one hand, and an equally growing demand for meeting the ever-escalating greed as expressed by the conspicuous consumption of the rich, on the other, have adverse effects on environment. The first results from a rapid growth of population and the second from the so-called "modern civilization".

At the global level the International Geosphere and Biosphere Project monitors the environmental change. The International Social

Science Council studies the human dimensions of environmental change. The Social Science Project has emerged from the realization that all changes in natural environment which concern humanity today have their roots in human behaviour. Such efforts will help assess science and technology from human and social angles also.

We need today a sort of technology which improves productivity while at the same time helps protect the natural environment. India is known for traditional methods of sustainable agriculture. The Himalayan slopes were not devoid of vegetation. Every clearing was converted into terraces for cultivation. Thus the natural slope after deforestation was modified to stop erosion. Prof Indra Pal mentioned how a similar pattern emerged in the Rajasthan desert. Now when we come to the current efforts, we find that there is a wide gap between both traditional and modern technologies but neither is being used adequately.

A good example of waste disposal system eminently suited to Indian conditions is *Sulabh Sauchalaya.* We know that the waste of one activity can become the basic resource of another. Even human excreta is being used as raw material for biogas in China. Urban garbage can be used to generate power instead of being dumped to pollute the land, water and air. Industrial waste also can be recycled. When we come to the social aspects of technology, we have to give weightage to the ethical dimensions. There is inequitable access to resources at all levels, individual to global. There is a tendency to grab the resources wherever they are. The greed for resources has gripped the whole of humanity. We must have all the data in this regard to improve the situation.

For the collection, storage and retrieval of data at the world level, there is global resource data base; we should have a similar facility at a National, State, and even District level.

In my view, the natural endowments should be counted as assets; they enter the national income account. If we start doing this, we will be more alert about the environment.

Ecology, Equity and Development

H Ramachandran

I would like to say that the anthropocentric issues of ecology are highly relevant to environment. We cannot, therefore, divorce economic issues from environmental issues. It would be rather difficult to implement an environmental code which goes against private economic interests. I give you a few illustrations.

There is a fair amount of evidence that shifting cultivation is environmentally undesirable. But if we work out the economics of it, we find that the returns to farmers from this cultivation are higher than from settled cultivation. The problem arises because of conflicts between private and public costs.

Recently I read a book entitled *Economics and Ecological Crisis*. On reading through the book, I felt that a better title would have been, "Ecology and the Crisis in Economics". The essential difference I am drawing is that the current methods of measurement in economics are incapable of handling ecological sensitivities. There are tools to measure economic changes, but we do not have adequate tools to measure ecological changes. To impute monetary values to ecological resources is absolutely unacceptable from the environmental point of view. Nevertheless, our academic discussions continue to empute market determined prices and work out costs and benefits and then justify one technology and one solution over the other. Thus I see the need for a different academic ethics when we deal with ecological and environmental issues.

Last year (1991), we submitted a report to the government of Karnataka on the Impact of Investment in Coastal Karnataka. One of the colleagues who handled environmental issues came out with recommendations about compensation for displaced persons. In the penultimate meeting with the government before the report was finalized, we were advised not to elaborate on the issue of displacement of people and compensation.

My colleague did not agree with this advice. He leaked the report to the Press, hoping that someone will take up the question. In one sense, the leak was unethical; but from the environmental point of view, it was desirable. Here the academic ethics and environmental ethics perhaps come in conflict with each other. This kind of dilemma I have seen earlier also. All voluntary agencies and the NGOs face such problems when they take up the issue of environmental degradation.

There are instances where the NGOs are not without vested interests. When the land being submerged belongs to big farmers, the shouting will be more. If it belongs to the poor, small and marginal farmers, the shouting will be less. These kinds of issues linked to the less privileged are either drowned in discussions or submerged in seminars and symposia.

I make two hypotheses for our consideration. First, while the environmental problems and production technologies have been addressed at various fora, the distributional aspects and their environmental implications have not received due attention. A technology policy that is environmentally benign must necessarily have to be socially benign. Thus whether we approach the issue from the point of view of social desirability or from that of environmental desirability, we reach a common point of convergence.

Whether our area of interest lies in social justice, welfare, economic growth or anything else, it has to be in tune with environmentally desirable policies and programmes.

People and Development

B M Shukla

In my view the solution to environmental problems does not lie in the governmental superstructure alone. It lies in the minds and hearts of the people. Gandhi represented that mind and heart of the common man of not only India but of the whole world. To say that Gandhi is dead is also to say that Truth is dead; Non-violence is dead; Human Brotherhood is dead; and Unity of the Universe is dead. Gandhi represented in his lieftime much of what Indian culture stands for. Then, in that case, Indian culture is also dead. The environmental ethics that we are looking for is definitely contained in Indian culture and in other ancient cultures. Gandhi summarized all this in one sentence: "There is enough on this earth to meet every one's need, but not every one's greed." I would like to remind Prof Dikshit that no matter what we get from science and technology, so long as the greed is not removed from the minds of the people, there will be no peace.

I am a student of Chemistry, but I do dare say that we scientists are not necessarily the custodians of all scientific knowledge. There is a lot of science inherent in the way the people interact with nature. There is science even in the prayer. Perhaps that science is still beyond the reach of our thinking and imagination. The common man knows more about the local flora and fauna than we do. We must make use of that knowledge also. We must be prepared to learn from people, as Gandhi said. But it is also true that there is a dimension of scientific knowledge that the common man either does not know or, if he knows, he does not know that he knows. It is the modern science which has developed it during the last 200 years.

We must use this latest knowledge derived from the great traditions of indigenous peoples the world over to solve our environmental problems. This is what Gandhi also wanted.

Finally, let me add that spirituality is a great cleanser of social dirt and garbage. The present-day environmental and ecological problems are in way the wastes produced by science and technology. You can lift this waste and put somewhere else by using new technologies. But if you want a society with only that amount and type of waste which our system can recycle, then you have no choice but to fall back on spirituality and then on to environmental ethics.

A Ghafoor Ghaznawi

It would be a formidable task for me to summarize the very extended and involved discussion we had this afternoon. What is clear from the discussion is that we should not try to juxtapose economy and ecology. The choice is not between development and environmental *serenity*. The choice lies in the judicious balance between the two.

Secondly, we should not think that we have inherited the planet earth from our parents; we should think that we have borrowed it from our children. Such a thought will encourage us to leave this planet at least in the same condition in which we received it, if not better. Our aim should be to have a worth living place and worth living life. If we can achieve it by reviving ethical values, or by evolving new ethical values, let us try it.

The ethical components of environmental education are deeply rooted in the cultures and great traditions and religious. These values can be used in alliance with the imperatives of science and technology to create a set of new values. We have to re-enforce the traditional approaches accepting and modifying them wherever necessary to the needs and challenges of contemporary civilization.

11

VALEDICTORY ADDRESS

11

VALEDICTORY ADDRESS

Shri K C Pant

Former Minister of Human Resource Development, India and Chairman Finance Commission of India

Friends, I am very grateful to Prof Misra for giving me the opportunity to come to Gandhi Bhavan which in my view is the most appropriate place to discuss environmental ethics. I firmly believe that Gandhiji is as relevant today as he was during the freedom struggle. What he stood for is relevant not only to India but to the whole world. One can, of course, argue about the details of his philosophy, approaches and strategies he used, whether one will use the same methods he used in his time. But if we look at the message of his life, it is not only relevant, but also a must for peace and harmony and, therefore, today's younger generation must be brought into contact with the essence of Gandhi's philosophy and thought.

If there is anything that can sustain this country in its quest for self-respect, self-reliance and self-esteem, it is the Gandhian approach, along with, of course, the opportunities which have emerged from modern science and technology. But I would insist that our whole approach to development must be harmonized with Gandhiji's concept of *Sarvodaya;* that while we integrate our economy in the world economy, we must understand that no nation can be built by others. We have to build our nation ourselves. And that self-reliance and self-help must be the central point of our endeavour. That is derived from Gandhiji's message to us. After all, he made the people of India fearless. He generated self-confidence in the people. Today, I think it is this fearlessness and self-confidence and faith in the future which are the essential ingredients for the progress of this country. That is why I said that Gandhi Bhavan is the right place for the Symposium.

I am here on the last day of the Symposium. I see that a great deal has been covered during the last three days. The session I attended discussed the cultural roots of environmental ethics. Even at the risk of repetition, let me say that the ancient Indian thought propagated respect for all forms of life on the earth. It believed in the oneness of life. And this has been one of the continuing and sustained currents of Indian culture.

This has been a part of Jainism wherein non-violence has been introduced as a natural corrolary to this basic approach. Belief in rebirth further confirms the need to respect other forms of life because you never know what form of life you had before, and in what you will be born, hereafter.

These are the elements of our mental make-up which subconsciously direct us to do something or not to do something. These then create a sort of collective subconsciousness which, in the context of a nation, can be termed as national subconsciousness. This subconsciousness often cuts across religions, castes and creed boundaries and creates an urge in the whole populace to act and react in a particular way.

When I was a child, I remember, we were told that it was a sin to break a green branch of a tree. I remember a festival in the Kumaon hills to celebrate greenery called *Hariyala*. It was to create a sense of respect towards plants and trees and also to encourage people to plant more trees. Several trees were worshipped. It was a means to preserve and protect those trees. I am told that all such trees have great ecological and medicinal importance. And so, I fully agree that where we can draw on our own experiences, our own cultural values and our own roots, we will stand on firmer grounds and we would be able to transmit those ideas in a more natural manner to our younger generation today.

When scientific truths are given cultural backing, they are easily accepted and assimilated. Why not we do this then, particularly when we see so clear a convergence in what science says and what Indian culture says on ecological and environmental matters. I feel this is a fit subject for research also. At the elementary level of education, we must draw upon the local experiences and local environment to make children understand the need to preserve the nature. The classroom education needs to be combined with outdoor real-life education.

Earlier, children in rural environment learnt in the natural way: how the trees grow, how the animals give birth, how the birds live and survive, how the crops grow, etc. But now the children in the upper class Delhi think that wheat bread grows in the refrigerator. Sometime back, I visited a primary school in a remote area Pithoragarh district of U.P. Himalaya. The headmaster of the school had a personal collection of the flora and fauna of that area and had drawn sketches and put us photographs of various plants and animals of that area. Let me tell you that is the best effort in this direction I have ever seen in a primary school. This is something that is a must to resurrect our education from the rut it has fallen in.

I fully support the Symposium conclusion that unless environment becomes a concern of the people and unless environmental movement becomes a people's movement, nothing substantial can be achieved. Governments do have a role to play. But the way governmental system is organized and structured, it can at best act as an enabling and catalytic force. It is clear from our day-to-day experience. Governments arrange planting of trees but not many of the saplings survive. People are not involved. Governments can do many things because they have the resources, but they cannot force people to do something they are not interested or involved in. And, therefore, the sooner we involve the people directly or through NGOs, the better it would be. We do not have time to waste. Ecological problems are already very acute. We should act soon.

Prof Misra's letter mentioned that the conclusions of this Symposium will go as an input to UNCED being held at Rio in June through the Commonwealth Human Ecology Council, London. In my view, for the purposes of that conference we, as a developing country, must make our position very clear. We have always accepted our responsibilities as a member of the United Nations. Those responsibilities enjoin on us to evolve environmental policies which are in the interest of the people of this country and the world as a whole. We are a part of the world community and we recognize that the environmental problems that the world is facing today are very serious. Whether we are responsible for them or not, we cannot escape the consequences of what has happened so far and the stage we have reached. When we look at the ingredients of the imbalances that have been created between man and his environment, we find that the non-renewable mineral and energy resources have come to exhaustion. The petroleum on which all of us depend for a variety of needs is marked by sky-rocketing prices and dwindling supplies. How long would it be available, we do not know. Perhaps not for too long.

We know that in Saudi Arabia and other Gulf countries, there is still a large reserve, but India has not found anything after Bombay High, to become self-sufficient in oil. And that alone will cost so much foreign exchange in the years to come that it is going to be a very serious problem for us.

I mentioned this because non-renewable resources can be conserved only if we really believe that the future of humanity is in our hands. We must not just fed for ourselves. We should not exhaust the earth. We must leave to the future generations adequate resources so that the progress of human civilization is not marred by a precipitous fall. It is in this time perspective that we should see the use of non-renewable resources.

Think of how we have treated the animal life. Hunting for pleasure, for ivory, for skins, for oil, etc., has eliminated many species, and many more are on the verge of extinction. Commercial value has cast aside all other values. Leave aside the ethical question of whether animals have as much right to exist and survive as human beings, can there be a different religious and cultural approach in this matter? People of all religions are concerned about it. I do feel that today all sensitive people in the world accept that every form of life must be preserved and development must conform to some of the basic biological imperatives. There must be regulation in the economic and commercial exploitation of all forms of life. We accept that fishing is necessary, because for a lot of people it is a source of livelihood and nutrition. We cannot say that there will be no fishing. That is not a practical proposition . But if we say that fishing should be done in such a manner that it does not exhaust the fishing resources for the future, then we are talking of a sound ecology and economy and no one should object to it. We must evolve methods which make fishing a sustainable enterprize.

We have removed the plant cover from many areas. Where should the people ago? We have built dams submerging large areas of forests and agriculture. Let us build dams if they are needed, but let us also see that proper habitat is provided to animals affected by submergence; let us put that much land under forests somewhere else at the same time we build the dams; and also let us rehabilitate the people uprooted from the project area in a manner that they are better off than before. What we have done so far in our country is not enough. We must do much more for the animal life, as well as for the poor.

Similarly, in respect of plant life, there is need to do something rather urgently. In olden times in India every village had a sort of virgin forest area, not very large, but sufficiently large to allow a variety of medicinal plants and herbs to survive. Necessary medicines were derived from these forests. They also functioned as habitat for small animals. Now, we do not have even a trace of these forests.

Those who have studied Kautilya's *Arthshastra* written more than 2,000 years ago know how it was obligatory on the part of the State to preserve and plant trees of various kinds. It is amazing that our country known for philosophies, religions and thoughts also had great scholars like Kautilya who went into the administrative details of the preservation of the environment. But in spite of all this, forest cover has come down to about 12 per cent even though the revenue records still say it is 18 or 20 per cent, as against the ecological need of 33 per cent of the total land area of the country.

Because of the loss of trees, the top soil is being lost. Soil is the product of hundreds of years of mutual interaction among various forces of nature. It is washed off within a decade or so carrying with it our capacity to evolve sustainable agriculture. It takes a very long time to rebuild the soil. We have to take note of it. Let us not wait for others to give company. Even if the rest of the world is not bothered about it, let us do it on our own.

The Club of Rome looked into these questions and warned of the dangers ahead. Many other reports have come up since then. They all point to a grim scenario and call upon us to adopt more environment friendly life and development styles. Fortunately, hundreds of NGOs have come up all over the world and in Germany a Green Party has already made a mark. They are the people who agitate against the nuclear waste being dumped in other countries or deep in the sea. There were agitations in Japan, New Zeland and the Pacific Islands against nuclear waste disposal. Similar movements were launched against whaling.

I am mentioning these movements because the people who have gathered here are also already or potential environmental activists. In the West, there is growing opinion that women should not wear firs of animals; crocodile and snake bags, shoes and belts have been banned. These reflect a heightened sense of concern for animals. We should be appreciative of tigers, lions, elephants and other forms of life in our own country. Same applies to plants. We must develop gene banks of the species which are on the verge of extinction. We should accept willingly some of these new concepts which are very much in conformity with our cultural values.

We have some of the world's best rain forests in Kerala, West Bengal, Andaman and Nicobar Islands, etc. The Government of India is trying to see that these virgin forests are not destroyed or even harmed. But unless there is a favourable public opinion and the right type of education of the people, these efforts will have only limited success.

Prof B M Shukla mentioned two areas which are causing great concern to us today. These are ozone layer depletion and Green House Effect. In both the cases discussions are on to fix responsibility. Such discussions are often unrealistic, because all of us are going to suffer irrespective of who is responsible for it. The development countries like India and China have large populations. If we go by the total emission of CFC, carbon monoxide or carbon dioxide, then certainly India and China with large populations would show up large volumes. But if we go by the per capita consumption of energy, we find that India is far behind the developed countries. And so the per capita contribution of carbon dioxide, etc., which creates the Green House Effect is very much more by the developed countries than by countries like India and China. If you consider the ozone layer depletion, the aerial spraying is very high in the industrialized countries not only in absolute but also in per capita term.

So the Question is: What can we do about it? As the country develops, the energy consumption also will go up. All the countries which are developed today are energy intensive. This intensity declines in post-industrial societies. India has just entered the energy intensive industrialization. We have not yet reached the peak. Unless we use more energy, we cannot solve our problems of low productivity, mass unemployment and underdevelopment in general. Even ordinary people cooking their food or lighting their houses, or using transport need energy. And so, I would very strongly urge that UNCED take note of the needs of the developing countries and give realistic targets. And if the industrialized countries of the North do not help us with environment friendly technologies, then we cannot really do much to contribute to environmental reconstruction.

We, in this country, have millions of people below the poverty line. The poor will not like to suffer further in the name of ecological balance and environmental cleanliness. Poor people have waited for generations for a better tomorrow. To ask them for another bout of waiting in the name of environment will not only be unfair, and unpractical but also unethical and against the basic tannets of Indian culture. All of us should rebel against any such immature idea.

In 1972, at Stockholm, India made a signal contribution by saying that poverty is the biggest polluter in the developing countries. Without food, water, housing, and other basic needs environmental protection has no meaning. If we do not take up poverty eradication and alleviation programmes in all the developing countries seriously, a great number of people will be deprived of the opportunities to live a full span of life. This aspect has to be re-emphasized at Rio de Janeiro.

Now I come to the aspect that is most intriguing in the sense that we know its gravity but we do not have the courage to do anything. That is population growth. The Indian National Congress under the leadership of Mrs Indira Gandhi took a serious note of it and evolved imaginative programmes to tackle this problem. But its political fallout was disastrous for her. Since then, no political party in India has taken it up seriously. It behoves us all to do something sooner than later. Let me use this forum to urge the intelligentia of this country to put enough pressure on the decision makers to respond positively and with a sense of urgency to the dangers ahead. India has made a lot of progress since Independence. But much of the gains has gone to serve a newcomers leaving our poverty base intact. We cannot go like this any longer. Population control is a must.

Coming to the question of the future and what would UNCED bring out, I must say that the success of the conference will hinge on whether or not the richer countries are prepared to share the responsibilities and are ready to provide adequate resources and appropriate technologies to enable the developing countries to participate in this worldwide movement to save the Earth from disaster. The countries which have gained from polluting the environment must now come forward to shell out part of the gain to depollute the environment.

The developing countries do not have the money to create or buy substitutes for a variety of pollution abatement devices. I think there is good reason why the developed countries should come forward to help the developing countries in a global effort to evolve strategies for depolluting the earth. For example, substitutes for aereal sprays which are CFC free have been developed now. But they will cost a lot. Poor countries cannot afford them. Therefore, for the common good, the developing countries should be helped.

But instead of helping, the Dunkel proposal on patent attempts to restrict the manoeuvrability of the developing countries to do anything meaningful. On the issue of Intellectual Property Rights, they want the patent laws to be so changed that access of the poorer countries to new technology would be further curtailed. There should be a selective approach to encourage the developing countries to join hands with the developed countries in this venture.

I do not think we should try to ape the consumerist culture of the developed countries. And if the problem of ecology and environment is to be faced on a long-term and sustainable basis, then we must find a substitute to the consumerist culture which is spreading all over the world rapidly. Every American family has two cars; and to sustain the automobile industry, the cars must be changed frequently. If every Indian and Chinese family had two cars, the Green House Effect would have already flooded many countries out of existence. So let us not follow these examples. But at the same time we can no longer tolerate a world in which there is so wide a disparity in living standards. And this message must be conveyed through UNCED. If it goes home, then the dialogue can begin soon to resolve various outstanding issues. Such a dialogue will help us to collectively find an answer to the environmental questions which are bothering us and which need urgent attention to ensure the survival of mankind through sustainable development of our economy and society.

Ladies and gentlemen, these are a few ideas I thought I should share with you. I once again thank Prof Misra for inviting me to join you this evening. I am confident that the conclusions you arrive at will help us in charting out policies and strategies which will make this Earth a better place to live on.

12

CONCLUDING REMARKS

Professor R P Misra

12

CONCLUDING REMARKS

R P Misra

Friends, with the closing address by Pantji, the deliberations of the Symposium come to an end. But the problems of ecology and environment will continue to haunt us. They cannot be resolved in a day. We need continuing dialogue like the one we had during the last three days. We need a feeling of shared responsibilities. Above all we need action. I wish we could continue the dialogue of cultures for some time more to arrive at a better judgment on policies and programmes of action. But we know that it is just not possible. So we regretfully disperse with the fond hope and resolve to continue to think about the recommendations we are making and to try in whatever capacity we are to follow them.

Let me now summarize the consensus we have arrived at. This consensus is the basis for the recommendations of the Symposium.

Defining Environmental Ethics

Ethics is that branch of philosophy which discriminates between right and wrong and translates values into options for action. From a large number of options open to man, he chooses only a few because of the limitations imposed by a variety of external and internal factors. One cannot disobey the laws of Nature or the State without retribution. Transgression of ethical principles, however, invites only social disapproval. It is not always followed by punishment.

This being the nature of ethics, it is never perfectly realized in any society individually or collectively. Yet at the same time no society can claim stability and orderly progress unless its members adhere to ethical provisions which substantially reduce uncertainties in human relations.

Ethical values whose transgression invites social unrest and chaos are usually encoded into formal laws. Laws are formally adopted, institutionalized and specifically sanctioned by society. There is, however, a vast social space which is no covered by laws. It is the moral and ethical values which guide the people in this space.

Environmental ethics is an internalized code of behaviour which regulates man's relationship with nature. Moral limitations on human behaviour in relation to the natural environment are needed because some forms of human behaviour are destructive of the whole or a component of the total environment. Since man is a natural being to begin with and his survival and development is very much linked to nature, he should use nature only to the extent that his life is not endangered. Environmental ethics ensures a healthy man-nature relationship.

People act as they perceive and think. And the perception and thinking of people is largely determined by the values they cherish. Man's attitude to nature and his habitat is, therefore, greatly influenced by ethical values. These values, however, are not universal, nor are their translation into action universally the same. While looking for a national or global code of environmental ethics, one has, therefore, to begin from local communities. These communities are the building blocks of far larger cultural realms or traditions.

Each local community develops a code of environmental ethics most appropriate to its natural conditions and genius. These codes keep a judicious balance between ecology and economy. In the past, these communities evolved sustainable agriculture even though at a low level of productivity. How can a similar system of agriculture, for example, be developed with a higher level of productivity forms the crux of the problem.

The advent of the Industrial Revolution as a driving force of the Cartesian worldview set in motion a new pattern of man-nature relationship. From the old pattern of man living in harmony with nature, man tried to control and subdue nature. Nature became are source for exploitation. The barriers of distance were annihilated, thanks to the steam engine, to lay the foundation of a global society; knowledge of the physical world multiplied fast, thanks to new inventions and discoveries, to capsule time beyond human comprehension; and freedom which Renaissance and Enlightenment promoted to free man of bondage of nature and traditions, triggered off unprecedented exploitation of man by man. The sword and the cross joined hands to create colonial empires of global dimensions.

In due course, the empires died, thanks to the inner contradictions inherent to this predatory concept, but the exploitation of nature to the point of ecodisaster on the one hand and division of world into rich and poor, on the other, continues even today.

The ethical values that once maintained ecological balance and helped people share their joys and sorrows are now thoroughly eroded. Individualism, materialism and scientism without balancing elements of humanism and spiritualism have given rise to a series of ecological and social problems. Some of these problems are too serious to be ignored or even postponed. They call for the reversal of current trends.

The reversal of trends would not take place in a day. It requires readjustment in all aspects of life-economy, ecology, science, technology and religion. This is, however, not a simple task. Any restructuring is painful; total overhauling means discarding of the Cartesian worldview and adoption of a humanistic worldview based on scientific knowledge and ethical principles.

Environmental ethics is an integral part of ethics in general. Clearly an environmental ethic is needed to curb these human abuses of nature now more than ever before as the human power to destory nature is now greater than ever. With the emergence of a global industrial human culture, indeed, the human impact on the environment has so increased in force, intensity and ubiquity that under the worst possible scenario imaginable—the nuclear holocaust—people may destory the environment (at least as we know it) along with themselves. Short of this cataclysmic and apocalyptic event, the global ecosystem may gradually be degraded to the extent that the forms of life presently existing (including human beings) are no longer adapted to radically changed conditions.

Rationale for Environmental Ethics

It is now clear that human actions which have direct destructive effects on the environment often have unsuspected deleterious effects on human beings. Deforestation, for example, not only destroys the forest ecosystem but also has climatological, hydrological and erosional effects in the area of deforestation as well as elsewhere, adversely affecting the welfare of the people, and making a few very rich. Deforestation is, therefore, unethical.

The above approach has important implications for human rights. All human beings have the right to live as they wish. Any interference without their consent is unethical. These rights can be extended to other forms of life, although with some riders.

The second approach is the moralist one. All human beings are morally accountable for the impact their actions, direct or indirect, have on non-human elements of the environment. While the human rights approach would not censure man for cutting virgin forests where no human beings live, the moralist approach would, because some of the plant and animal species may be extinct because of deforestation. Man has no moral right to do that.

Environmental ethic may, therefore, be classified into two types,

(1) anthropocentric, and (2) biocentric. The former is secular and more acceptable in the West. Judeo-Christian tradition gives man the stewardship of the earth. Dominion of man over the rest of the creation forms the basis of a powerful environmental ethic.

According to this tradition, God created a world which He declared to be good; He handed it over in the trust of man, whom alone He created in His own image, to dress and keep. Man, therefore, alone among creatures has not only the right to use the creation but also the responsibility to conserve and perhaps even enhance the capacities of other creatures for self-expression. This concept is specially useful in checking the extinction of plant and animal species for nothing would violate the spirit of stewardship than the wanton and careless destruction of biodiversity.

The biocentric ethics is rooted in Eastern philosophies Hinduism, Buddhism, Confucianism and Taoism. In these traditions man was not given the stewardship of the creation. He is an integral element of "Life". All living and even non-living things have an equal right to exist. Man being the supreme creation must use his wisdom to ensure that sanctity of nature and life is preserved.

Man has to take quite a lot from the environment for his survival and development. But this taking must be like taking milk from cow. The cow's life is sacred; it must be preserved so that it continues to give milk. Nature is likened to mother, and man-nature relationship as umbilical. Such a concept puts many limits on Man's use of nature.

There is an undercurrent of revulsion against the Western culture in the East. Most of the ills of society are often ascribed to the Western influence. There is an urge to revive the old values, the values which led this part of the world to give rise to some of the greatest civilizations of the world. The mad rush for all that is Western and apparently new is resisted, even though unsuccessfully. The chances for the revival of the traditional environmental ethical values are now, therefore, brighter than ever before.

Another trend that can help revive and review the traditional environmental ethical values, both in the West and the East, in the North and the South, is the growing search for identity at the local and regional levels in almost all nation-states. This identity in the olden times was based on spacebound local communities. Each community and locality was *proud* of its history, its uniqueness, and its traditions. There was a sense of pride in the local environmental quality. The new trend towards a distinct identity and communitarianism within globalism can be successfully used to rebuild and preserve the environment.

These traditional values have to be debated in the context of existing scientific knowledge. Whether one accepts it are not, scientific rationality and scientific knowledge are great assets of the modern society. The revival of traditional environmental values cannot be at the cost of scientific rationality. They must support each other, not work against each other.

Towards Global Environmental Ethic

Can there be a Global Environmental Ethic? Often, the answer is, no. But if the world is becoming a global village,it must also have global ethics. The strength of a village is the ethics-based community bond. The strength of the global village cannot, therefore, be without global ethics. Today we are witnessing the merger of nations into a global community of humans. The global village has become a visible reality because of the United Nations, multinational corporations, growing economic interdependence, and shortening of distances by new transportation and communication technologies. Environmental problems do not stop at national borders, nor do the natural disasters. There is a growing subscription to the Eastern view that all living species have the right to exist and no nation has the privilege to destroy any of them within or outside its borders. And hence the possibility of a global environmental ethic.

One can look at the Global Environmental Ethic from two interrelated perspectives : people's and government's. The people's environmental ethics concentrates on the behaviour of individuals and small groups like families and communities. The government oriented ethics concerns the behaviour of the governments. For example, the people all over the world are fighting an unprecedented direct to indirect violence against other living species. A large number of NGOs are involved in this exercise. They are fighting against the World Bank, national governments, and business and industry. Their struggle is no always based on legal rights; it is based on moral and ethical rights. The values which form the bases for these struggles have to be strengthened. Incidently, the people involved in these struggled do not accept national boarders in matters of ecology and environment. They are truly global in approach.

There is a need to evolve environmenatl ethics for governments too. No government should be allowed to take shelter under national sovereignty in matters of ecology, environment and human rights. In an inequitous world that we have to day, it is not easy to evolved such a global code of conduct; but as a first step, its need must be recognized and efforts must commence to codify them. In fact, in the immediate future, global ethics will continue to depend on national governments for adoption and implementation through various laws and statutes.

Unless serious efforts are made at both the NGO and government levels, evolution of environmental ethics at the global level as well as at the local and national levels would take too long a time to avoid and forestall ecodisasters.

Revival and development of environmental ethics at various levels, individual to global, calls for two simultaneous steps. The first involves farming of a serious of laws at the global, national and sub-national levels to protect nature from wanton destruction; and the second, to develop new educational programmes and technologies to sensitize people from the early childhood with environmental and ecological principles and imperatives.

Several countries have passed environmental laws to check, air, water, land and noise pollution. Some countries have regulations regarding the use of certain chemical fertilizers, insecticides and medicines as they have proved to be harmful to ecology and human health. Consumer protection laws, food adulteration laws, air, water, and noise pollution laws are all now almost universal.

These legal measures, however, suffer from three major infirmities. Firstly, many of the laws go only part of the way. For example, putting a statutory warning on cigarette packets that "cigarette smoking is injurious to health" does not help much. There has to be restriction on smoking not only in public places but also privately. Society provides medical services to all at subsidized rates. Those who indulge in smoking put undue pressure on this system. They get sick and also induce sickness in the others. Smoking should, therefore, be banned altogether.

Such half-hearted measures are found in other environmental areas too. Air pollution cause by motor vehicles, for example, is tolerated almost everywhere either because they are not implemented. Take the case of deforestation. There are laws against deforestation in one form or the other almost in every country. But, felling of trees, continues unabated.

At the global level, there are treaties and agreements which have the force or law, but they leave enough loopholes for the powerful nations to violate them. For example, the NPT puts restrictions on the non-nuclear States to acquire nuclear capabilities,but leaves the nuclear powers free to test and develop new weapons. One can cite several other international conventions which suffer from similar drawbacks.

The second weakness of the legal approach lies in its negative orientation. Most laws tend to restrict people and organizations from causing damage to environment. But, laws that encourage and reward people who follow environmentally ethical behaviour are few, if any. It is, therefore, not enough only to curb the tendency of the people to misuse land and destroy soil. Those who use land according to the ecological principles should be rewarded.

The third weakness of the national statuets and international agreements, treaties and conventions is their pro-rich bias. They tend to favour the powerful and the rich who also happen to be the polluters. We have case of NPT, biodiversity, IPR, transfer of technology, international trade, and many other areas where poor nations are being pressurized to accept conditionalities which would not be in their long-term interest; while those who are powerful are able to enforce the *status quo* or further strengthen their position. The above situation holds good in the case of national laws too. Laws against deforestation, pouching of wild animals, pollution of rivers and lakes and unscientific use of land, etc., are so framed that they affect the poor adversely. While slum clearance is a good idea, where should the slum dwellers go? While a check on feeling of trees is fine, what should the forest dwellers do for fuel wood, etc? While large dams and large factories do push up national economic growth, what should the persons displaced by the dams do? These issues have not yet been adequately taken note of by the national laws.

This brings us to the gap between State Laws and Human Rights. All human rights cannot be covered by formal laws. And hence everything that is illegal is not necessarity immoral or unethical. Laws in fact leave a lot of functional space which, unless filled by ethics, will create a highly chaotic social situation.

Environmental Education

Education, in its broadest sense, is a means to create and inculcate ethics. Environmental education can thus play a very crucial role in evolving, revitalizing, reviving, and ensuring environmentally sensitive human behaviour.

Education is a process of learning. It is life-long and is not limited to any age group, institution, or aspect of life. The day a child is born, it starts learning. In fact,m according to the earliest as well as latest findings, a child learns even when in the womb. In the early part of his life, he gets a unified and integrated learning environment. As years pass, the unity of knowledge is fractured and he sees the world in parts and sections. By the time he enters the high school, he sees no relationship between nature, man and culture. Soon he turns out to be a physicist, an engineer, a historian, or a poet. As a physicist, he sees a nature as matter; as engineer he finds the way to use it; as a historian he studies human progress as the record of man's mastery over nature; and as a poet he composes the songs of this victory or laments over what man has done to himself and to the earth he inhabits. Each one sees the world through a tunnel of his own making; and each one tries to make the best by ignoring the legitimate claims of others. The end result is a world full of pollution of air, water, land, mind and society.

Education from early childhood shapes the personality of man who as a multitude forms the society local, national and global. The current environmental crisis can, to a great extent, be attributed to an educational process which is neglectful of the role natural environment plays in human survival and development.

Three dimensions of environmental education deserve urgent attention. These are: formal and informal instruction including cultural diffusion by oral or self-learning processes; mass media; and philosophical and scientific research on human and natural ecology. The three are very closely related. Unless ecology and environment get a right place in the world views of various cultures and societies, neither mass media, nor schooling can become effective instruments of environmental sensitivity. This necessitates incorporation of ecological and environmental studies at all levels of education.

Unlversity Education

At present universities and other higher educational centres are geared to the Western worldview. Analytical approaches require breaking down of the whole into bits. Modern research is engrossed in a depth study of parts or segments. It has led to great scientific discoveries. But the relations of the parts with the whole, and among the parts, have not been studied and understood adequately to foresee the injurious effects of new knowledge and discovery.

Here lies one of the basic causes of ecological imbalance and environmental pollution. Analysis without synthesis gives depth of a limited segment of the whole. Application of this knowledge produces many unintended sides effects. Ecology and environment call for research on interrelations among the various forces of nature and human activities as much as is depth analysis of each force and activity. Such approaches are, however, not accepted in the discipline-based university system. A scientist is not concerned with ethics; and ethics is not concerned with science. Physical sciences and biological sciences did not have a meeting ground until very recently. Biophysics and biochemistry have come to acquire meaning only in recent years. And hence the question of science and religion having a common ground does not arise.

According to Vedic sciences, the functioning of an atom is the same as the functioning of the universe and all that comes in between. What is essence it says is that there is a general theory that explains the functioning of all phenomena whether physical, biological, or human. Disciplines concerned only with apart of the totality cannot produce a general theory and hence are partial and inadequate as a system of knowledge.

Once the university education gets out of its disciplinary non-overlapping circles of concern, it will produce knowledge without adverse, direct or indirect, effects. The process has already begun. Fusing of various physical, biological and human sciences has opened up new opportunities. As this intensifies further, a far more ecology and environment sensitive knowledge would emerge.

The environmental impact assessment methodology is still very imperfect. A lot more research is needed within various disciplines and also across disciplines. Training of engineers, scientists and social scientists in EEA is urgently needed. Universities have to be assisted and encouraged to take up these tasks.

Primary and Secondary Education

School provides the first opportunity for a child to come out of the influence of the family and parents, and to lean about the world beyond through others: children and teachers. What the child learns has much to do with school environment and the quality of teachers. The curriculum is important but far more important is what the teacher does and the image of the teacher the child fixes in his mind.

At this level, students must be taught about ecology and the role of each individual and group in maintaining ecological balance. They should be taught about environmental pollution and the cause thereof. They should be taken of factory sites; areas with soil erosion, deforestation, polluted rivers, etc., so that they can feel and experience what they read in books and hear from teachers. The extracurricular activities must include trekking in hilly, forested, and waterfront areas. Students should be taught to appreciate nature and healthy environments. Field work to collect local plant specimens and classroom study of each plant community must form part of instruction. Each discipline has its own environmental and ecological dimension. Course outline and details of each subject must end up with environmental implications.

Teacher education, therefore, needs reorientation. Ethical and environmental dimensions have to be brought in forcefully in teacher education.

Education at Home

What a child learns during the first five years of his life is most crucial for its personality. In this period, the parents, particularly the mother, are the teachers and home and its surroundings the environment. In a society where parents have no time to show their love and affection to children; where they are too busy in the mundane affairs to spend some time with them; where the child is trained to consume milk and bread produced in refrigerators, people cannot be environment friendly. Similarly, children, who grow in poverty and pestilence, in disharmony and conflict situations, cannot be environment friendly.

Children in the developed countries grow in a culture which takes pride in annihilating nature, in cutting trees, in hunting wild life, in the banality of individualism, and in equating life with matter. And those in poor developing countries grow in a culture which preys on all that nature provides, physical or biological, to meet their basic human needs. And these very children, when they enter adulthood and manhood have no compunction in destroying nature for short-term gains.

So only such persons would be sensitive to ecology and environment and would be guided by ethical values as protect, conserve and enrich nature while taking life-supporting elements from it; as are brought up in an environment which is wholesome, which is not marked by deprivation or consumerism, which is sensitive and sympathetic to nature and fellow human beings.

Perhaps the most important element of this environment is the mother. She is the best teacher one can think of. The love and affection the mother has for her children is unparalleled. None else can match her. Hence, the education of the mother with respect to ecology and environment is a crucial factor in the upbringing of children.

The condition of women in general and mothers in particular is not good at all in the developing countries. Illiteracy, ignorance, poor or no maternal care, poverty and deprivation haunt them day and night. Only very few women are lucky enough to lead a decent life. This situation has serious implications for children. Most of them grow under utter neglect. Many of them do not survive infancy. To expect these children in adulthood to be humane and sensitive to ecology and environment is to expect too much.

In the developed parts of the world, lots of children grow up in the care of single parents. Those who are brought up in healthy homes are never given the knowledge and experience of how the other half of the world lives and what role nature plays in their life. They are obsessed with more goods, comfort and luxury.

Children brought up in broken homes and in poor homes often turn out to be violent towards nature as well as other fellow beings. Any meaningful environmental education programme must, therefore, be integrated with overall socio-economic development programmes with a thrust on women's development and women's education. This naturally takes us beyond the purview of environmental education perse, but then that is the way things are and should be. Education and development have too close to be treated independently when it comes to mannature relationship.

Mass Media

Mass media, print or electronic, play a very important educative role. They carry the message to all those who are neither in universities nor in schools. In fact, they bring to the common man dimensions and implications of various ecological and environmental problems. They are the moulders of public opinion, which in turn, moulds the opinion of decision-makers at local, regional, nations, and global levels. Electronic media (both audio and video) have given new and better opportunities for moulding public opinion and to sensitize people in favour of ecological balance and environmental reconstruction. Here is a very powerful tool to reach even the remotest corners of the world at a relatively low cost. They can also reach even those who are not formally educated.

For the mass media to become effective channels of environmental education, they need reorientation from the source of entertainment to the source of entertainment and education. They also need adequate trained manpower not only to handle the gadgets but also, and more importantly, to make the software. At present, we do not have enough journalists who take interest in ecology and environment. There are quite a number who look for sensational news, but very few who present a scientific and illustrative document for general readers. Courses in journalism in universities do not consider ecology as a major theme. The pattern is, however, changing but only slowly.

With the advent of radio, TV, satellite imageries, computers, etc., the situation is a little utter in the electronic media, but deep ecology is still at the margin. We have all the means at our disposal, but we do not yet have adequate and appropriate skills of and knowledge for presentation, nor do we have enough people who can handle environmental issues and create progrmmes for various target groups, such as children, students in junior highs, colleges and universities, adults, professionals, businessmen, politicians, etc.

The use of mass media and various educational technologies in much of the world, particularly the developing world, is extremely limited. Not even 0.001 per cent of the primary schools in India, for example, have newspapers, radio or television sets. How can the children benefit form these programmes. This applies to adults too to a great extent.

The developed world can help countries like India in the field of electronic media, on both the hardware and software sides. What we need is a massive environmental education programme through mass media all over the world. People are too ignorant about the critical man-nature relationship in the North because of ignorance and poverty. While it will take time to remove arrogance and poverty, something can be done to remove ignorance sooner. Mass media are important means to achieve this goal.

Vote of Thanks

Let me now thank you all for joining us in this Symposium and contributing immensely to make this meeting a success.

I know the arrangements for the Symposium left much to be desired. Fortunately, the nature was kind. You had a naturally air-conditioned weather. It was not cold; nor it was hot. It was warm though from the human angle. We had a wonderful opportunity to exchange our views and to learn from each other. But it was not the meeting of the minds alone; it was more importantly the meeting of hearts. We had lots of ideas and abundant love. So the inconveniences, if any, I hope, were drowned in the flood of love and affection.

I am personally grateful to Pantji for having joined us this evening. I have known him for the last one decade. He is one of the politicians in

India who has occupied senior ministerial positions with untainted grace, dignity and humility. In leaders like him lie the hopes of the future. I am equally thankful to Mr Kamal Nath, Minister for Environment and Forests, for inaugurating the seminar. Mrs Zena Daysh has been a source of inspiration to a lot of people all over the world in matters of Human Ecology. It is said that each time she visits a place, a new Human Ecology association or institution comes into being. I do not know what would emerge from her this visit to Delhi but I do know that even before she thought of coming over here, a new voluntary organization called Sustainable Development Foundation was set up. SDF is one of the sponsors of this Symposium.

Zena Daysh is a person who talks less, but does more. What she has been able to achieve singlehanded, many well-established institutions cannot. The other day, I asked Zena about the philosophy of her life. She spontaneously said "Love". With her presence her Love pervaded all round.

I must mention the role of the UNESCO. Through Dr A Ghafoor Ghaznawi for making this symposium a success. I thank Prof Gangrade for the contribution he made and the time he gave to the Symposium. I must say a few words about my friend from Jaipur, Prof Unnithan. We have a lot of similarities, including the style of the beard both in cut and colour. That helps us not only to look alike but also to think alike. We are beholden to him for cooperation and help extended to us.

I am thankful to you all, ladies and gentlemen, for being here with us.

APPENDICES

APPENDIX A

PROGRAMME OF THE SYMPOSIUM

March 1, 1992

9.00 a.m.	—	**Registration**
9.45-11.00 a.m.	—	**Inaugural Session** Chairman : Shri B N Pande
11.00-1.00 p.m.	—	**Session I** *Philosophical Foundations of Environmental Ethics.* Chairman: Mrs Zena Daysh Prof Vidya Niwas Misra
2.00-3.50 p.m.	—	**Session II** *Environmental Ethics and the Emerging World Order.* Chairman: Prof B M Shukla Dr Anil Agrawal
4.00-5.30 p.m.	—	**Session III** *Law and Environmental Ethics.* Chairman: Prof Upendra Baxi Prof P S Sangal

March 2, 1992

9.00-11.00 a.m.	—	**Session IV** *Tribal and Rural Experiential Knowledge and Environmental Ethics.* Chairman: Prof B K Roy Burman Prof R R Singh

11.15-1.30 p.m. — **Session V**

Culture, Religion and Environmental Ethics
Chairman: Prof K D Gangrade
Prof D N Chaturvedi

2.30-4.00 p.m. — **Session VI**

Population Growth and Environmental Ethics

Chairman: Prof Ashish Bose
Prof Indra Pal

4.00-5.00 p.m. — **Session VII**

Science, Technology and Environmental Ethics

Chairman: Dr B R Nanda
Dr S K Sinha

March 3, 1992

9.00-11.00 a.m. — **Session VIII**

Education and Environmental Ethics

Chairman: Dr Ghafoor Ghaznawi
Prof T K N Unnithan

11.15-2.00 p.m. — **Session IX**

Development Styles and Environmental Ethics

Chairman: Prof G S Nathawat
Prof G L Pandit

3.00-5.00 p.m. — **Closing Session**

Chairman: Prof S K Aggarwal
Prof T K N Unnithan

Voledictory Address: Shri K C Pant

Concluding Remarks: Prof R P Misra

APPENDIX B

List of Participants

1. **Shri AGRAWAL, ANIL**
   Centre for Science and Environment
   F-6, Kailash Colony, New Delhi-110 048

2. **Prof AGRAWAL, S K**
   Secretary, Association of Indian Universities
   Kotla Road, New Delhi-110 002

3. **Dr AGGARWAL, S K**
   Department of Geography
   Delhi University, Delhi-110 007

4. **Shri ADHIKARI, N S**
   Gandhi Smarak Nidhi
   Rajghat, New Delhi-110 001

5. **ANTO, FATHER**
   Sadbhavana, 259-F, Pocket-I
   Mayur Vihar, Phase-II, Delhi-110 009

6. **Dr ATHREYA, M B**
   A-28, Chittaranjan Park
   New Delhi-110 019

7. **Prof BALI, T N**
   Department of Hindi
   University of Delhi, Delhi-110 007

8. **Prof BAXI, UPENDRA**
   Vice-Chancellor
   University of Delhi, Delhi- 110 007

9. **Shri BHARATI, RAM SINGH**
   337, Gali Prakash
   Teliwada, Delhi-110 006

10. **Shri BHARADWAJ, A B**
    Gandhi-in-Action
    B-29, Wazirabad
    Bhajanpura, New Delhi-110 003

11. **Dr BHARATI, SHASHI**
Department of Hindi
Satyawati College,
University of Delhi
Ashok Vihar, Delhi-110 052

12. **Prof BOSE, ASHISH**
I-1777, Chittaranjan Park
New Delhi-110 019

13. **Prof CHATURVEDI, D N**
Former Vice-Chancellor
Kashi Vidyapeeth, Varanasi.

14. **Shri CHITKARA, AMRIT PRAKASH**
Municipal Corporation Primary School
Roop Nagar, Delhi-110 007

15. **Mrs DAYSH, ZENA**
Vice-Chairman
Commonwealth Human Ecology Council
57/58 Stanhope Gardens SW7 5RF
London

16. **Mrs DESHPANDE, NIRMALA**
Chairman, Harijan Sevak Sangh
Delhi-110 009

17. **Prof DIKSHIT, K R**
L-4 Ganesh Khind
University of Poona, Pune

18. **Dr DUA, N K**
19/190, Sarai Basti
Delhi-110 035

19. **Dr DUBEY, R N**
Lecturer in Geography
Bhim Rao Ambedkar College
Delhi University, Delhi-110 031

20. **Prof GANGRADE, K D**
156, Vaishali
Pitampura, Delhi-110 034

21. **Shri GEORGE, T**
    Joint Secretary
    Ministry of Environment and Forests
    New Delhi-110 003

22. **Dr GHAZNAWI, A GHAFOOR**
    Environmental Education Unit
    UNESCO, Paris

23. **Dr GOSWAMI, S C**
    Shri Ram College of Commerce
    Delhi University, Delhi-110 007

24. **Dr HÖSLE, N**
    Department of Philosophy
    University of Turingen
    Pnebmunallee 1, D 8100

25. **Shri JENAMANI, SURVENDRA**
    Researcher,
    Department of Geography
    University of Delhi, Delhi-110 007

26. **Shri JHA, AMAR NATH**
    Swami Shradhanand College
    Alipur Road, Delhi-110 036

27. **Dr KAMAL, RAKA**
    Sakaar 35, House Colony
    Civil Lines, Jaipur-79327

28. **Shri KAMAL NATH**
    Minister of Environment and Forests
    Govt. of India, New Delhi-110 003

29. **Prof KAPOOR, KAMAL DEVI**
    Osmania University
    City College for Women
    Hyderabad

30. **Dr KAPOOR, RAVI**
    Department of Sociology
    Jamia Millia Islamia
    New Delhi-110 025

31. **Prof KAUSHIK, SUSHILA**
Department of Political Science
University of Delhi, Delhi-110 007

32. **Sri KUMAR, A**
A-16, Pandrik Vihar
Pitampura, Delhi-110 034

33. **Shri KUMAR, AWADHESH**
Social Worker
Pitampura, Delhi-110 034

34. **Shri KUMAR, SATISH**
Department of Law
University of Rajasthan, Jaipur

35. **Prof MACHWE, M K**
Deptt. of Physics
Delhi University, Delhi-110 007

36. **Prof MISRA, H N**
Dept. of Geography
Himachal Pradesh University
Shimla

37. **Prof MISRA, ANIL DUTTA**
Dept. of Political Science
Delhi University, Delhi-110 007

38. **Dr. MISRA, RAJ KUMAR**
422/559, Misra Bhawan
Bahadur Ganj, Allahabad-110 003

39. **Prof MISRA, R P**
Director
Gandhi Bhavan 32, Chhatra Marg
University of Delhi, Delhi-110 007

40. **Prof MISRA, VIDYA NIWAS**
Vice-Chancellor
Kashi Vidyapeeth, Varanasi

41. **Dr MOHANTY, M K**
Dept. of Philosophy
Delhi University, Delhi-110 007

42. **Shri MOR, TARA CHAND**
Haryana Gram Vikas Samiti
H.No. 1592, Sector-15
Sonepat, (Harayana)

43. **Shri NANDA, B R**
S-174, Panchshila Park
New Delhi-110 017

44. **Dr NARANG, SUDESH**
29, Miranda House
Delhi University, Delhi-110 007

45. **Shri NATH, AKHAY**
Research Activist Sahayog
86/B.D.Estate, Delhi-110 054

46. **Prof NATHAWAT, G S**
Indira Gandhi Centre for Human Ecology
University of Rajasthan, Jaipur, (Rajasthan)

47. **Prof NEELAMEGHAM, S**
Faculty of Management Studies
University of Delhi, Delhi-110 007

48. **Prof PAL, INDIRA**
A-14 Anita Colony
Jaipur, (Rajasthan)

49. **Dr PANDE, B N**
Vice-Chairman
Gandhi Smriti and Darshan Samiti
5, Tees January Marg, New Delhi-110 011

50. **Prof PANDEY, K K**
Ayurvedic and Unani Tibia College
Karol Bagh, New Delhi

51. **Shri PANDE, VINOD KUMAR**
Kuswapur
Pratapgarh, (U.P.)

52. **Prof PANDIT, G L**
Dept. of Philosophy
Delhi University, Delhi-110 007

53. **Shri PANT, K C**
Chairman, Finance Commission
Govt. of India, New Delhi

54. **Dr PATANJALI, P C**
Principal
Dr Bhim Rao Ambedkar College
Delhi University, Delhi

55. **Prof POSER, HANS**
Institute for Philosophy and History of Sciences
Technical University of Berlin, Berlin

56. **Shri PRAKASH, ARUN**
Dept. of Geography
Delhi University, Delhi-110 007

57. **Dr RAMACHANDRAN, H**
Reader, Dept. of Geography
University of Delhi, Delhi-110 007

58. **Dr RAUTH, MADHUMITA**
215, Sabarmati Hostel
INU Campus, New Delhi-110 067

59. **Prof RAZA, JAFAR**
Dept. of Urdu
University of Allahabad
14-B, Bank Road, Allahabad

60. **Dr RUKMINI, T S**
Principal, Miranda House
University of Delhi, Delhi-110 007

61. **Shri ROY, BIR CHANDRA**
Secondary Teacher
High School; Bharat Khand (P.O.)
Distt. Khagaria, (Bihar)

62. **Dr RUSTAGI, URMILA**
Miranda House
University of Delhi, Delhi-110 007

63. **Prof ROY BURMAN, B K**
Centre for the Study of Developing Societies
Rajpur Road, New Delhi.

64. **Prof SANGAL, P S**
Campus Law Centre
Delhi University, Delhi-110 007

65. **Dr. SAXENA, BIMLA**
30 Teachers Flats
Miranda House, Delhi-110 007

66. **Dr SHARMA, JAYA**
225, Sabarmati Hostel
J N U Campus, New Delhi-110 067

67. **SHARMA, R C**
Reader and Head
Dept. of Geography
M D University, Rohtak, Haryana.

68. **Dr SHASTRI, SATISH**
Department of Law
University of Rajasthan
Jaipur-302 004.

69. **Dr SHUKLA, B**
Post Box 2189
Delhi-110 007

70. **Prof SHUKLA, BENI MADHO**
Former Vice-Chancellor
Gorakhpur University
35, Nanda Nagar, Varanasi

71. **Dr SHUKLA, J R**
Dept. of Geography
K S Saket Post-Graduate College
Faizabad (U.P.)

72. **Shri SHUKLA, KALIKA PRASAD**
Nachiketa Gram Vikas Pratisthan
Pratapgarh, (U.P.)

73. **Dr SINGH, C P**
Dept. of Geography
University of Delhi
Delhi-110 007

74. **Dr SINGH, C B**
Gandhi Bhavan
University of Delhi, Delhi-110 007

75. **Dr SINGH, MAN**
Lecturer in Chemistry
Deshbandhu College, Kalkaji
New Delhi-110 019

76. **Dr SINGH, R B**
Dept. of Geography
University of Delhi, Delhi-110 007

77. **Prof SINGH, R R**
Delhi School of Social Work
University of Delhi, Delhi-110 007

78. **Dr SINGH, SAVITA**
Sr Research Officer
D-26, Gulmohar Park, New Delhi

79. **Shri SINGH, SURENDRA BAHADUR**
Freedom Fighter
Pirthiganj, Pratapgarh (U.P)

80. **Dr SINGH, SUJEET KUMAR**
Campus Law Centre
University of Delhi, Delhi-110 007

81. **SINHA, RAJEEV KUMAR**
Asstt. Professor
Indira Gandhi Centre for Human Ecology
University of Rajasthan, Jaipur, (Rajasthan)

82. **Prof SINHA, S K**
Director
Indian Agricultural Research Institute
New Delhi

83. **Dr THAKUR, LAXMI**
Assistant Professor
Indira Gandhi Centre for Human Ecology
Environmental and Population Studies
University of Rajasthan
Jaipur

84. Shri TIWARI, SUDHAKAR PRASAD
Advocate
Pratapgarh (U.P.)

85. Shri TIWARI, RAJ NARAYAN
Nachiketa Gram Vikas Pratisthan
Pratapgarh (U.P.)

86. Mrs UNNITHAN, G J
P-3, University Quarters
Jaipur (Rajasthan)

87. Prof UNNITHAN, T K N
P-3, University Quarters
Jaipur (Rajasthan)

88. Shri VERMA, V P
Department of Philosophy
Delhi University
Delhi-110 007

89. Shri VASUDEVAN, N
Gandhi Book House
Gandhi Smarak Nidhi
New Delhi-110 001

90. Prof YADAV, HIRA LAL
Dept. of Geography
631-A Mohaddipur
University of Gorakhpur
Gorakhpur

Note: The above list does not include the names of student participants, nor does it include those who attended the Symposium only partly and did not make any oral or written presentation.

Index